INTRODUCTION TO BIODETERIORATION

SECOND EDITION

This book provides an introduction to biodeterioration – the attack on man-made materials by living organisms. The authors outline the principles involved, as well as the ways in which such damage can be controlled and prevented. A wide range of organisms are covered (including bacteria, fungi, algae, lichens, insects and other invertebrates, birds, mammals, and plants), and the following types of biodeterioration are discussed:

- natural materials such as foodstuffs, wood, metal, stone, cellulose, and leather
- synthetic products such as paint, adhesives, and plastics
- structures and systems such as buildings, monuments, transport systems, and vehicles.

In this updated and expanded second edition, new molecular and genetic techniques are included, and regulatory, environmental, and safety issues are emphasized. This book is suitable not only for biologists, but also for those in industry, commerce, and local government who are concerned with the preservation and conservation of a wide range of materials of economic or cultural importance.

Dennis Allsopp is formerly of CAB International Mycological Institute, where he was Head of the Culture Collection and Industrial Services Division, and latterly CAB International Biodiversity Coordinator. He coedited *Smith's Introduction to Industrial Mycology* and coauthored the first edition of *Introduction to Biodeterioration*.

Kenneth J. Seal is the Technical Director of Thor Specialities in the UK, which develops and produces a range of biocides and textile chemicals. He was previously Head of the Biodeterioration Division of the Cranfield Institute of Technology Biotechnology Centre and co-authored the first edition of *Introduction to Biodeterioration*.

Christine C. Gaylarde works in the Department of Biophysics and the Microbial Resources Centre at the Federal University of Rio Grande do Sul, Brazil.

Introduction to Biodeterioration

SECOND EDITION

Dennis Allsopp

Kenneth J. Seal

Christine C. Gaylarde

CAMBRIDGE
UNIVERSITY PRESS

PUBLISHED BY THE PRESS SYNDICATE OF THE UNIVERSITY OF CAMBRIDGE
The Pitt Building, Trumpington Street, Cambridge, United Kingdom

CAMBRIDGE UNIVERSITY PRESS
The Edinburgh Building, Cambridge CB2 2RU, UK
40 West 20th Street, New York, NY 10011-4211, USA
477 Williamstown Road, Port Melbourne, VIC 3207, Australia
Ruiz de Alarcón 13, 28014 Madrid, Spain
Dock House, The Waterfront, Cape Town 8001, South Africa

http://www.cambridge.org

First published 2004

Printed in the United States of America

Typefaces Utopia 9.5/13 pt. and Helvetica Neue   *System* LaTeX 2_ε [TB]

A catalog record for this book is available from the British Library.

Library of Congress Cataloging-in-Publication Data
Allsopp, D.
    Introduction to biodeterioration / Dennis Allsopp, Kenneth Seal, Christine
C. Gaylarde. – 2nd ed.
        p.   cm.
    Includes bibliographical references.
    ISBN 0-521-82135-5 – ISBN 0-521-52887-9 (pb)
    1. Materials – Biodeterioration.   2. Biodegradation.   I. Seal, Kenneth J.   II. Gaylarde,
Christine C.   III. Title
    TA418.74.A39   2003
    620.1′1223 – dc22

                                                                                    2003060350

ISBN 0 521 82135 5 hardback
ISBN 0 521 52887 9 paperback

Dedications

FIRST EDITION

Dedicated to the memory of our friend and colleague
Professor Thomas Alan Oxley
1910–1983

SECOND EDITION

This edition is dedicated to the memory of
Mr Arthur David Baynes-Cope
1928–2002
A gentleman, friend, colleague, and scholar who did
so much to advance the role of science in the study of
biodeterioration problems in museums and libraries.

Acknowledgements – Expert Contributors

The authors are most grateful to
Dr David Pinniger and Dr Adrian Meyer
for their revisions of and contributions to the insect
and rodent sections of this edition.

Contents

Preface to the second edition

Since the publication of the first edition in 1986, the field of biodeterioration has expanded, evolved, and inevitably become more complex. New techniques in molecular biology are now routinely employed, and regulations on biocide use have been refined and made more stringent in moves towards integrated pest management and environmental protection.

Despite these changes, the aim of this book and the approach we have employed have remained constant. This book is intended as a basic introduction to biodeterioration, which is the attack on materials of economic significance by living organisms. The subject is huge, and this remains a very small book. The authors have attempted to cover most of the important areas; to this end, there are now three of us and we have enlisted the help of other specialists in certain fields. Nevertheless, some readers will find their particular interest poorly represented. The approaches to the subject of biodeterioration vary widely, depending on the specialist area. A pest-control officer will view the subject differently from a bacteriologist, a biocide manufacturer differently from a museum conservator. We hope that our approach will be of help and interest to a wide range of readers who are not yet specialists in particular aspects of the topic and also to many in areas in which biodeterioration is infrequently considered. A basic level of understanding of biological systems is assumed. Those readers who do not have such an understanding are referred to the more general texts which are included in the reading lists for each chapter.

On the advice of many of our colleagues involved in teaching, and in response to comments on the first edition, we have kept most of the text free from literature references, concentrating them instead in reading lists which give a guide to the most useful and easily obtainable texts.

Many people have helped in the production of this book, and our attitudes and approach to it have been shaped by many colleagues and friends worldwide over the past 30 or more years. We thank them all. The production and revision of a small book on such a huge subject are not easy; once again we open ourselves to the inevitable charges of incompleteness and imbalance. Nevertheless, we hope that this new edition will contribute to greater awareness of and interest in the topic, particularly for students and others beginning to explore this important and diverse field.

2003

D.A.

Porton, UK

K.J.S

Marple Bridge, UK

C.C.G.

Porto Alegre, Brazil

INTRODUCTION TO BIODETERIORATION

SECOND EDITION

1

Introduction

DEFINITIONS

What *is* biodeterioration? The word has only been in use for about 40 years, but describes processes which have affected humankind ever since we began to possess and use materials. Many branches of science and technology either do not need or do not have an accepted definition in common use. We are all happy to think we know what physics is, but we have yet to agree what exactly constitutes biotechnology. Within biodeterioration we are fortunate to have a definition which was quickly accepted when first proposed by H. J. Hueck.

Hueck (1965, 1968) defined biodeterioration as '*any undesirable change in the properties of a material caused by the vital activities of organisms*'. Another term in common use is that of biodegradation. Although no formal definition has general acceptance, it may be useful to think of biodegradation as being '*the harnessing, by man, of the decay abilities of organisms to render a waste material more useful or acceptable*'. Both definitions involve humankind, in a negative or harmful way in the case of biodeterioration and in a positive or useful way in the case of biodegradation as defined here. Both definitions also involve materials. Materials are any form of matter, with the exception of living organisms, which are used by humankind. All materials (and processes) will have an intrinsic value and thus there is an important economic dimension to biodeterioration. This book seeks to introduce some of the interactions between living organisms and humankind's materials through the disciplines of environmental biology, materials science, and ecology.

The interaction among people, their materials, and living organisms has been recognized, if not fully understood, for a very long time. Some of the interactions are intimate, as in the case of gut parasites, body lice, or fungal infections of the skin, hair, and nails. Others are not so close, as in the case of commensal pests, such as rats, thriving as a result of human activities. The relationships can be complex; a commensal rat may pass on more intimate insect parasites or bacterial and viral pathogens to a human population. In the case of pathogens there is usually a clear reaction from the living tissue affected, involving an attempt to limit or remove the pathogen, whereas in the case of deteriogens of materials there is no such reaction. This is one of the fundamental differences which separates biodeterioration of materials from plant or animal pathology. In both biodeterioration and pathology, damage is caused in some way, and it is this damaging involvement of organisms which is considered here.

Many of early humankind's materials were derived from plants and animals, with minimum processing. Such materials were particularly vulnerable to attack by organisms, and biodeterioration may have played an important role in the ways in which early civilizations were able to develop and spread. What we now know as the Middle East is regarded by many as the 'cradle of civilization'. The climate is suited to the easy storage of grain, and such an advantage may have speeded human development in this region. Early control of biodeterioration utilized the basic principles which still hold good today. Food was either eaten fresh before any form of deterioration could occur, dried to minimize the growth of microorganisms, or physically protected from insects and rodents by use of sealed jars and bins. Salt and spices were used as preservatives. In the non-food area, early examples of control included the use of burning sulfur as a general fumigant and the use of copper sheeting on ships' hulls to provide a physical barrier to boring organisms.

In the world today, there is a bewildering range of materials. Many are complex and much changed from the original raw materials from which they were derived. New environments are, for good or bad, being exploited; roads and cities are being built in rain forests, structures are erected in seas and oceans for oil exploration and extraction, and high-rise office and apartment blocks take advantage of new building techniques. These new materials, their uses, and environments present biological problems, and together with the enormous range of organisms in the environment it is useful to classify the basic types of biodeterioration which can occur. It should be noted, however, that any such classification scheme is artificial, and one organism may cause more than one type of biodeterioration (see Figure 1.1)

Figure 1.1. Mould growth and staining on cotton wrapping of glass-fibre pipe insulation from a UK hospital. The mould is not only breaking down the cellulose of the cotton, but also causing disfigurement. Spores released into the air may also cause problems in a hospital environment. Photo: Dr K. J. Seal

PHYSICAL OR MECHANICAL BIODETERIORATION

In this instance, the organism quite simply disrupts or distorts the material by growth or movement and does not use it as a food source. There are few, if any, serious examples of such damage caused by microorganisms, but one which might be quoted is the expansion of microbial masses between rock layers, leading to spalling of the surface. Examples caused by higher organisms include the cracking of underground pipes by tree roots, the gnawing of electrical cables, cinder blocks, plasterboard, and wood by rodents, and bird strikes on aircraft. This latter example illustrates the point that biodeterioration is not necessarily caused by any 'conscious' process of the organism.

FOULING OR SOILING (AESTHETIC BIODETERIORATION)

Here, the objection is simply to the presence of an organism or its dead body, excreta, or metabolic products. Dead insects, moult cases, or

Table 1.1. Some problems associated with the presence of biofilms on materials

Biofilm location	Effects
Teeth	Tooth decay, caries
Medical implants	Antibiotic-resistant infection, weakening of implant material
Heat-exchanger tubes	Reduced heat transfer
Pipes carrying water or other liquids	Reduced flow, increased resistance to flow
Cooling towers	Reduced performance, degradation of material, provision of reservoir for pathogens (e.g., *Legionella*)
Drinking water distribution systems	Decreased water quality, increased treatment costs, health risks
Probes and sensors	Reduced efficiency
Ships' hulls	Increased fuel costs
Building materials	Reduced durability, discolouration
Food-processing equipment	Source of contamination, degradation of material, increased cleaning costs
Screens and filters	Loss of efficiency
Oil industry pipelines	Blockage and corrosion

droppings, even if in some cases not particularly harmful in themselves, can render foodstuffs unsaleable, especially in packages in developed countries. Microorganisms, especially fungi and algae, can be found growing on otherwise undamaged materials, utilizing surface dirt and detritus, but nevertheless detracting from the value or acceptability of the material. The classic example here is the dark fungal colonies growing on damp soap and skin residues on plastic shower curtains. The performance of the material is not affected, but the growth creates a generally unacceptable appearance. Many fungi may also release soluble or insoluble pigments and also a range of other metabolites which discolour on ageing.

Fouling can be more serious and transcend the category of purely aesthetic damage, in that a physical function can be impaired. The extra drag on ships caused by accumulations of weed and invertebrates on the hull can increase fuel consumption dramatically, and extra tidal stresses on marine structures such as oil rigs can be considerable.

In many cases, aesthetic biodeterioration is simply the presence of a surface layer of microorganisms and their products. These microbial layers are known as biofilms and are defined as surface accumulations of the organic products of biological activity. Generally, but not always, they include living microbial cells and, in this case, their presence may lead to the production of other classes of biodeterioration. Where the cells are in such close contact with the material, rather than merely dispersed in the surrounding environment, destructive cell activities are localized and concentrated and the resulting effects far more damaging. The best studied biofilms are those on the teeth, and the results of bacterial activities in these biofilms are well recognized by the layperson. Other damaging effects of biofilm presence are, however, less well understood. Some of these are shown in Table 1.1.

(BIO)CHEMICAL ASSIMILATORY BIODETERIORATION

This is probably the most easily understood form of biodeterioration. Quite simply, the organism is using the material as a food or energy source. Microbial enzymes breaking down cellulose, rats and insects eating stored grain, and insect larvae consuming stored fruit are all examples of this type of biodeterioration.

The consumption of human foodstuffs by deteriogenic organisms is in the main recognized and understood by most people, it is the variety of the microbial diet which can cause problems. The fact that materials such as hydrocarbon fuels, cutting oils, adhesives, sealants, textiles, and other 'non-food' items can be utilized by microorganisms is often not realized and can lead to delays in establishing the cause of problems.

(BIO)CHEMICAL DISSIMILATORY BIODETERIORATION

In this instance, a material suffers chemical damage, but not as a direct result of the intake of nutrients by the organism. Many organisms excrete waste products, including pigmented or acidic compounds, which can disfigure or damage materials. This type of biodeterioration often goes hand-in-hand with chemical assimilatory biodeterioration and biofilm development, and the effects may be difficult to differentiate.

THE RANGE OF DETERIOGENS

No special criteria exist which debar any organism from being an actual or potential deteriogen, except obligate parasites, which gain their nutrients from the living tissue of their host (living tissue is not a 'material'). Such parasites, however, may well be associated with deteriogens (e.g., the rat flea). The range of potential deteriogens is therefore huge, although with many organisms the deteriogenic effects are a minor incidental part of their activities as a whole.

CYCLING OF ELEMENTS

Within the biosphere, elements are constantly recycled. The reader will be familiar with the carbon and nitrogen cycles, and it is possible to propose cycles for other elements, although some (e.g., sodium) are present in the environment in such huge quantities as to make the effects of living organisms insignificant. Figure 1.2 shows a simplified carbon cycle with both biodeterioration and biodegradation taken into account. The carbon cycle is strongly biological, and its main feedstock is the carbon tied up in the cellulosic components of higher plants. The ultimate fate of all cellulosic materials used by humans is to be either burnt or broken down by organisms, and the prevention of biodeterioration is thus delaying the natural decay processes which would normally affect the materials if they had not been abstracted from the environment by people for their use. The biodegradation of wastes can be regarded as the opposite process: a hastening of breakdown of materials complementary to natural processes in the environment.

RECOGNITION AND COSTING OF BIODETERIORATION PROBLEMS

Cases of biodeterioration must first be recognized as such before any economic costs can be attributed to them. Some cases of damage to materials are obviously of biological origin. Holes gnawed by rats in woodwork, drains blocked by weed growth, and pigeon droppings disfiguring ledges of buildings are all obvious examples. Less certain are cases such as the appearance of fruiting bodies of fungi on rotten woodwork and mould growth on foodstuffs. There is obviously a problem, but not everyone recognizes such 'growths' as living organisms. Cases of microbial growth well mixed or dispersed in materials, such as bacteria in cutting fluids, fungi in hydrocarbon fuels, and the spotting of paint films by microfungi and algae, are less easy still to recognize and appreciate, owing to the small size of the organisms involved, their unfamiliarity, and the fact that the substrate is chemically very different from animal and human foodstuffs.

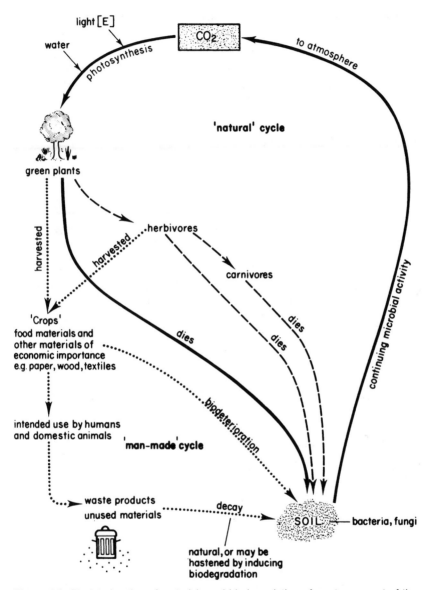

Figure 1.2. Biodeterioration of materials and biodegradation of wastes as part of the global carbon cycle.

Instances such as microbially induced corrosion of metals, in which separate physicochemical mechanisms may also be in play, can be even more difficult to recognize and quantify.

Even if damage to materials is recognized as being biological in origin, it may not be listed as such in routine reports, being simply classed with other types of losses as 'spoilage' or 'wastage'. Evidence of the true cost of biodeterioration is therefore often hidden or disguised. Recognized biodeterioration may be costed in a variety of ways.

Cost of prevention

Where materials are known to be at risk, preventative measures can be taken from the outset. These measures may be physical, such as drying or cooling, or chemical, such as the addition of a biocide or preservative. The costs of such measures can be assessed for individual products by a producer or user, and, on a more general level, the turnover of a 'prevention' industry or firm such as a biocide manufacturer can be used as an indicator of the cost of biodeterioration in a particular field.

Cost of replacement

The number of materials, especially in the more developed countries, which can be classed as 'cheap' is dwindling rapidly, but where low-cost materials are used, the replacement cost can be used as a guide to the cost of biodeterioration. All other considerations being equal, if the cost of replacement exceeds the cost of prevention, the wrong strategy is being employed.

Cost of remedial measures

Any decision to apply remedial measures to restore a material to near its original condition hinges on the cost and practicality of such work as opposed to that of replacement, always assuming that a replacement is available.

The best examples of true remedial work are those carried out on costly or unique items such as museum specimens and archival items, for which prevention or replacement is not possible. Such work is usually very costly, and true remedial work is uncommon on everyday materials.

Cost of litigation

Occasionally, the biodeterioration of a material will be due to neglect on behalf of one or more parties in the chain of production/supply leading to another party's seeking recompense through the law for loss of value. This can result in the employment of expert witnesses, analytical laboratories, and lawyers to find the cause of the problem, apportion blame, and decide on the level of compensation which may go far beyond the value of the lost goods.

In the life of some materials the costs of biodeterioration may come from all these basic areas, initial preventative measures, remedial measures during the use of the material, and eventual replacement. Anyone with wooden window frames will appreciate this. There is no single universal scheme for the costing of biodeterioration. However, there are certain exercises which can be carried out to reach very approximate global figures and at least show that biodeterioration can cost significant sums of money.

A broad calculation

1. Select types of materials with known susceptibility to biodeterioration. These would include timber, paints, adhesives, natural-fibre textiles, paper and packaging materials, and stored foodstuffs.
2. Estimate a percentage of loss which would be assumed to be reasonably attributable to biodeterioration. Some stored or in-transit foodstuffs in the tropics may have losses of around 90% of the crop. To arrive at a figure which is beyond argument, decide on 1% (faced with this problem, most people decide on 5%–20%). Is it not reasonable to say 1% of all timber used fails prematurely because of biological attack?
3. Find the annual value of goods produced, whether locally, nationally, or on a world scale.
4. Take 1% of this figure. However approximate this calculation is argued to be, a large sum of money is involved.

This figure gives an indication of the importance of prevention of biodeterioration.

In the following chapters, biodeterioration is divided into sections on materials in which relevant organisms and control measures are described.

Principles of control are covered in a separate chapter, as are recent developments in detection and research methods.

REFERENCES AND SUGGESTED READING

Allsopp, D. (1985). Biology and growth requirements of mould and other deteriogenic fungi. *J. Soc. Archivists*, **7**, 530–3.

Allsopp, D. and Gaylarde, C. C. (2002). *Heritage Biocare; Training Course Notes CD*. Archetype, London.

Eggins, H. O. W. and Allsopp, D. (1975). Biodeterioration and biodegradation by fungi. In *The Filamentous Fungi, Vol. 1, Industrial Mycology*, Smith, J.E. and Berry, D.R. (Eds). Arnold, London, pp. 301–19.

Eggins, H. O. W. and Oxley, T. A. (1980). Biodeterioration and biodegradation. *Int. Biodeterior. Bull.*, **16**, 53–6.

Florian, M.-L. E. (2002). *Fungal Facts*. Archetype, London.

Hueck, H. J. (1965). The biodeterioration of materials as part of hylobiology. *Mater. Org.*, **1**(1), 5–34.

Hueck, H. J. (1968). The biodeterioration of materials – an appraisal. In *Biodeterioration of Materials*. Eds. Walters, A.H. and Elphick, J.S. Elsevier, London, pp. 6–12.

Kirk, P. M., Cannon, P. F., David, J. C., and Stalpers, J. A. (Eds.). (2000). *Ainsworth and Bisby's Dictionary of the Fungi, 9th ed.* CAB International, Wallingford, Oxon. U.K.

Rose, A. H. (1981). *Microbial Biodeterioration*. Vol. 6 of Academic's Economic Biology Series. Academic Press, London.

2

Natural materials

The division of biodeterioration into topics is convenient but artificial. Many 'natural' or non-synthetic materials have specific major uses: for example, wood and stone are common building materials and are therefore considered in Chapter 4 as 'materials in use'. Where materials, such as wood, are dealt with elsewhere, only the basic mechanisms involved in their biodeterioration are considered in this chapter.

Cellulosic materials

Cellulose is the main structural component of plant cell walls and is probably the most abundant biological compound on Earth, the total world estimate being approximately 26.5×10^{10} tonnes. Cellulose is produced in vast quantities, and the fact that it is recycled relatively quickly emphasizes both its susceptibility to attack by organisms and the range of organisms able to do so and utilize the breakdown products. In green plants, cellulose is the main material found in the primary cell walls, which are thickened in structural and conductive tissues. These special tissues provide many of the plant fibres used by humans. Cellulose is also found in the higher fungi and algae. It occurs alone in the fungi and algae, but in higher plant groups it is frequently associated with lignin, which is much less readily biodegraded.

Cellulose is a linear polysaccharide made up of β-(1–4) linked D-glucopyranose residues (Figure 2.1), occasionally cross-linked to other similar chains by means of hydrogen bonding to produce microfibrils.

Lignin, however, although made of the same basic chemical elements as cellulose, namely carbon, hydrogen, and oxygen, is not built of sugar,

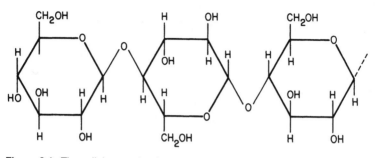

Figure 2.1. The cellulose molecule.

but of phenylpropane subunits, and is not a carbohydrate. It forms an irregular reticulate molecule rather than chains, with a molecular weight in the region of 100 000 or more, and its chemistry is very complex. Lignin protects cellulose from microbial attack, as shown by the greater resistance of sisal or jute fibres compared, say, with that of cotton. Cellulose occurs as the major constituent of wood (and thence, by processing, of paper and chipboard) and similarly as plant fibres used in textiles; for example, flax and jute are phloem fibres, sisal and manila are leaf fibres, and cotton is made up from seed fibres.

The susceptibility of cellulose to biological attack depends on the presence of suitable environmental conditions for colonization by organisms and also on the physical and chemical form of the cellulose, which will vary depending on the type of product being made. Physical breakup of cellulose fibres by grinding can increase their susceptibility. When paper is manufactured, the cellulose is delignified, yielding a more susceptible material. Native cellulose is mainly crystalline with some amorphous sites, and as it undergoes physical and chemical processing many more amorphous sites are created along the macromolecule. It is this change which makes cellulose itself more liable to attack. Much laboratory work has been done on highly modified amorphous cellulose, such as filter paper, cellulose powder, and even regenerated cellulose such as cellophane. The use of these easily handled 'pure' forms of cellulose is no doubt simpler for the laboratory worker, but the results obtained should be treated with caution if they are to be related to the decay of wood or straw, as the cellulose fractions within such materials will not be open to the same amount of attack.

Cellulose is often chemically modified in the production of derivative forms for particular uses. These modified and regenerated forms of cellulose have uses as textiles (rayon is regenerated cellulose), packaging films (such as cellophane), and photographic film (cellulose nitrate and acetate), and as thickeners, fillers, and extenders (e.g. carboxymethyl cellulose) in adhesives, foods, and emulsion paints (see Chapter 3). Many of these forms of cellulose are susceptible to biological attack; the thinning of emulsion paint in storage and the use of cellophane as a 'bait' in soil ecology studies are examples. In the past, attempts have been made to chemically modify the cellulose molecule (as opposed to the use of biocide treatments) to confer greater resistance to biological attack. Such techniques, however, do not appear to have been particularly satisfactory or economic in practice.

The breakdown mechanism of cellulose is not only an example of how microorganisms can degrade large molecules, natural or synthetic, but also how enzymes released by organisms can continue to break down materials even when the cells which produced them are removed or no longer alive. Free enzymes can be a particular problem in liquid industrial systems. All living organisms possess a range of intracellular enzymes, active within cells, which catalyse the various biochemical reactions involved in biosynthetic and energy-liberating processes. In active cells, the most important of these intracellular enzymes are continuously produced (constitutive enzymes). When a suitable substrate is available in the immediate environment of the cells they may produce other types of enzymes, known as inducible enzymes, which may be intracellular or extracellular and are produced only when the appropriate substrate is present. Extracellular enzymes are needed to degrade the substrate molecules to a size that the transport mechanisms of the cell membranes can take up for further breakdown and utilization within the cell.

Cellulose (the molecular weight of which varies from 200 000 to 400 000 or more depending on the source) is usually broken down to glucose through a number of intermediates by a system of several enzymes known as the cellulase complex. This complex comprises the following enzyme systems: C_1, (an exoglucanase), C_x, (an endoglucanase – also known as carboxymethylcellulase or CMCase), and a β-glucosidase (cellobiase). If excess glucose (the end product) is added to the system, the production of the cellulase enzymes complex is repressed, that is, the system is self-regulated by the amount of glucose present. It is important to know whether an organism is capable of C_1 activity, as the ability to utilize most laboratory cellulose media, which consist mainly of amorphous cellulose, indicates only C_x (CMCase) activity and is not indicative of an ability to degrade native cellulose, which has more crystalline components needing C_1 for initial breakdown.

Once liberated, extracellular enzymes are not dependent on the organism for their activity. Such an effect can be seen in the thinning of contaminated aqueous emulsion paints thickened with a cellulose ether such as carboxymethylcellulose in the form of a dispersed gel. The organisms can be suppressed or destroyed by addition of a biocide, but the enzymes may remain unaffected, continue to degrade the thickener, and cause complete loss of viscosity. Other extracellular enzymes such as pectinases, amylases, and proteases can act in similar ways: Indeed, this phenomenon is responsible for the activity of 'biological' washing powders, which contain proteases and lipases as an aid to dirt removal.

Some bacteria are known to be able to degrade cellulose. Cellulolytic bacteria such as *Cellulomonas* and *Cellovibrio* can be found on particular substrata, for example wood, but isolation onto laboratory media can be difficult. More work has been carried out on the actinomycetes which are able to colonize alkaline wastes from cellulose-processing plants, rapidly removing the cellulose fraction. Bacteria probably have their greatest effect on cellulose under very wet conditions (e.g., the 'ponding' of logs to increase permeability of wood to wood preservatives), but even when the extent of their role in cellulose decay is more fully understood and quantified their total effects may be small compared with those of the fungi. Other environmental conditions related to the breakdown of cellulose involve the availability of nutrients, one key factor being the carbon-to-nitrogen (C:N) ratio. In many types of wood, the C:N ratio is very high, from 350:1 up to 1250:1; this low availability of nitrogen severely limits the potential of fungi to attack the wood. Timber derived from the heartwood of trees and the refined forms of cellulose such as filter paper and processed cotton are often less susceptible to attack if a source of nitrogen is lacking. Some basidiomycetes, however, are able to cycle nitrogen translocated from moribund hyphal tissue behind the growing tips of their own mycelium. They are able to penetrate and attack heartwood, producing brown and white rots. With brown rots the lignin is largely left untouched, whereas white-rot fungi destroy both lignin and cellulose.

There has been increased interest over the past 20 years in the mechanisms of degradation of lignin, brought about by the realization that the enzymes involved are relatively non-specific and might be useful in bioremediation of environments polluted with recalcitrant aromatic chemicals.

These ligninases have been investigated for their ability to degrade pollutants such as textile dyes and pesticides. Most of the work has been done with the basidiomycete fungus *Phanaerochaete chrysosporium*, but there is intense research in progress on other fungi. The extracellular lignin-degrading enzymes were first discovered and characterized in *P. chrysosporium* in the 1980s. They include lignin peroxidases (LIPs) and manganese-dependent peroxidases (MnPs), as well as hydrogen peroxide-producing oxidases. Hydrogen peroxide is necessary for the action of the lignin-degrading enzymes. In other fungi, laccase, a blue, copper-containing phenol oxidase, may also be involved. Without the action of these enzymes, our planet would be covered with a thick layer of undegraded plant material and life as we know it would be impossible.

In the past, hardwood heartwood was used almost exclusively for building purposes. More recently, with greater demand for materials, the use of softwood/sapwood which has been insufficiently seasoned has resulted in the early onset of decay. During seasoning the nutrients in the sapwood slowly migrate to the surfaces of cut timbers where they can eventually be removed by further machining or washing.

Apart from the common major types of structural decay previously mentioned and discussed in more detail in Chapter 4, there are other significant types of fungal attack on wood. Soft rots of timber are forms of superficial decay which tend to occur in continuously damp or wet conditions. They are caused mainly by a range of microfungi (less robust than the basidiomycetes, which may explain the need for constant moisture) including ascomycetes such as *Chaetomium globosum*, an organism which has been found to be involved in serious cases of decay of wooden fillings in water cooling towers.

A variety of fungi may cause staining of timber, either by liberation of a pigment as in the case of *Chlorociboria aeruginascens*, which causes a green stain in oak (to such an extent that the stained wood is used as a veneer in decorative inlay work), or by the presence of dark fungal hyphae within the wood, as in the case of blue stain, caused in temperate regions by the ascomycete *Ceratocystis*. Other fungi are known to cause staining, for example, *Scytalidium* sp. and *Gliocladium virens*. Staining of wood, in particular blue stain, does not significantly alter the strength of the wood, but it does reduce its value. Staining fungi are usually controlled by the application of water-borne fungicides to sapwood timber immediately after felling. Wood chips and wood pulp can also be adversely affected by staining fungi.

Together with wood and textiles, paper is another major cellulosic product susceptible to attack by microorganisms. Attack may begin even before the paper is formed, pulpwood timber being attacked by a variety of basidiomycetes, which weaken the fibres. Pulp may be attacked by both basidiomycetes and a range of microfungi, leading to both decay and staining.

The papermaking process itself provides a wet, warm, and nutritious environment that is conducive to the formation of bacterial slimes, which cause clogging of papermakers' felts and irregularities and weaknesses in the finished product. Restrictions on the discharge of wastewater (known as white water) from papermills and the consequent recycling of water can also tend to increase the risk of microbial contamination. Biocides are regularly used in papermills to control the development of these slimes. Once made, paper has a low water content and the main microbial spoilage organisms are restricted to the fungi. When paper is made into books, the nutrient and ecological factors become more complex. A book contains a variety of materials: as well as paper there may be textile fibres, fillers, glues, board, and perhaps plastics or leather. All these have different nutrient and water-absorption qualities, and as books are used and handled, other materials, such as sweat and grease from the hands of the user and dirt from the environment, will accumulate over time. The conservation and preservation of books is a complex topic; they cannot be considered as simple assemblages of sheets of cellulose.

Apart from microorganisms, many other organisms (insects, earthworms, molluscs, and crustaceans) are able to attack cellulose in a variety of ways. Many insects are able to damage wood, using it as a nutrient source and for shelter and egg laying (see Chapter 4). The ways in which adult insects and/or their larvae digest wood fragments varies. Some appear to possess the ability to secrete cellulases directly within their alimentary tract, but some, for example, termites, have a complex symbiotic relationship with microorganisms, or protozoa and microorganisms, present in their gut, which secrete cellulases and enable the insect to utilize some of the breakdown products. Some insects bore into wood to obtain the cell contents, mainly sugars or starches, and utilize only these compounds, the cellulose cell walls not being digested. Some may utilize cellulose but also make use of cell contents and other nutrients, if available, to increase their rate of growth and development. The woodworm or furniture beetle (*Anobium punctatum*) develops much quicker in old-fashioned plywood than in plain timber, owing to the presence of animal glue in the former.

Molluscs and crustaceans in the marine environment are also able to bore or tunnel into cellulosic materials (usually wood), using it for harbourage in which to feed. Examples are filter feeders, such as the shipworm (*Teredo* spp.), or browsers on surface bacteria and algae, such as the gribble (*Limnoria* spp.).

WOOD IN THE MARINE ENVIRONMENT

Marine wood-borers

No account of the biodeterioration of structures would be in any way complete without mentioning marine wood-borers, such as the shipworm and the gribble, which have played a significant role in human endeavours and travels for hundreds of years. Wood introduced into the marine environment, in the form of ships, piers, and barriers, is produced at considerable expense and effort and always serves some purpose. Such wood is attacked by a range of marine organisms worldwide, and two common examples, one crustacean and one mollusc, are discussed here.

The gribble (*Limnoria lignorum*)

Limnoria is a common crustacean and can be found in most wooden pier and harbour piles which have stood for any length of time in the sea. Three species of the isopod genus *Limnoria* are found around UK shores, the commonest of which is *L. lignorum*, a small organism about 3 mm long and usually present in large numbers. Each animal has seven pairs of short legs ending in sharp claws, which enable it to grip the sides of its burrow. *Limnoria* bores into wood using mandibles which border its mouth. These mandibles are asymmetrical and pass over each other in a rasp-and-file action. A series of galleries are constructed below the surface of the wood, reminiscent of those made by some species of drywood termites. This riddled layer of wood is eventually eroded away by wave or current action, and a fresh, deeper layer is then exposed and open to further penetration and damage.

Each branch of the tunnel or gallery system harbours two gribbles, one male and one female. After mating, the female produces twenty to thirty large eggs, which she carries, like all isopods, in a brood pouch under the body. After hatching, the young, which are about one-fifth adult size, leave the female and immediately start boring their own tunnels leading off the parental domicile.

Limnoria may possess no cellulase enzymes of its own. In feeding it does ingest some wood, but it is thought that much of its food is obtained in the form of microscopic algae, bacteria and soft-rot fungi which occur in and on the wood. It seems that *Limnoria* may bore mainly for protection, as they have been found in seemingly non-nutritious and non-wood environments such as the insulation covering submarine cables.

Shipworm (*Teredo navalis*)

There also exist molluscan wood-borers, the most familiar being the shipworms. These specialized elongated molluscs were responsible for destroying Sir Francis Drake's ship, the *Golden Hind*, and in the 1730s threatened the Dutch polders by devouring the wooden dyke walls which had been used in the reclamation of land from the sea.

The commonest species of shipworm is *T. navalis*, which is larger than the gribble, growing up to 45 cm long. Detection of *Teredo* damage can be difficult, because the only external evidence of its presence in wood is a small hole. The body is long and worm-like, with the shells confined to the tip. *Teredo* produces a multitude of planktonic larvae, which on attachment to wood begin their metamorphosis into the adult form.

The reduced shells of *Teredo* are efficient cutting tools, possessing sharp ridges which rasp into the wood. There have been some interesting comparisons of the 'design' of these cutting surfaces with modern machine tools. Small plates (pallets) at the siphon end of the organism can close off the small external opening of the burrow, and the animal can survive in its burrow for several weeks if the wood is removed from water, a useful adaptation in tidal waters. Some cellulose digestion occurs, and planktonic food is also taken in through the siphons.

The other wood-borer worthy of mention is the wharfborer beetle *Nacerdes melanura*, which lives in waterlogged wood. It is commonly found in old buildings or river wharf pilings in fresh or brackish water. The large white larvae tunnel in the wood, and the adult beetle emerges above the waterline. Historic boat timbers have been seriously damaged during storage.

Control of marine borers

Metal sheathing of wood has been used in the past, a technique which also has applications in prevention of marine algae and barnacle fouling (hence the term 'copper-bottomed'). The main drawback with regard to marine borers is that the sheathing needs to be perfect and continuous, with no holes or gaps. This is almost impossible to achieve, and suitable malleable metals such as copper are both heavy and expensive. Alternatively, wood preservatives have been used, particularly copper–chrome arsenate (CCA) compounds, applied initially by pressure impregnation. Leaching of wood preservatives is a problem in the marine environment, and reimpregnation may be difficult or impossible. Even the use of explosives has been suggested, to kill the organisms by shock waves. The force needed to achieve this, however, is so great that the structure itself would inevitably sustain damage.

STORED FOOD

The material of primary concern to everyone is food. Ever since humans have collected food and stored it they have been concerned with its preservation from decay. The topic is vast, as diverse as the different types of food in the world, and only some basic examples can be given here. The diseases which affect crops while they are alive and growing fall into the sphere of plant pathology and will not be covered in this book. Our concern in biodeterioration is with post-harvest decay or spoilage, but as the organisms themselves do not always recognize the difference, there is, of course, some overlap; some 'field' fungi become 'storage' fungi once the crop is harvested.

A wide range of microorganisms can attack foodstuffs, which is hardly surprising, as human foods are essentially the same as the dead material which is recycled in the natural world. Post-harvest losses are at their greatest in tropical and sub-tropical climates because of the higher temperatures and also poorer control and storage techniques sometimes found in the less developed countries. Estimates of the magnitude of losses vary, and it is not always clear how great a part is played by biodeterioration: However, the figures give cause for concern. Generally losses of the more 'durable' crops such as grain, oilseeds, and legumes are mainly due to biological agencies, whereas the more perishable crops with a high moisture content (over 50%), such as soft fruit and salad vegetables, generally show greater losses that are due to mechanical damage during harvesting or handling or to extremes of temperature.

Post-harvest losses of durable crops in Africa, Asia, and Latin America have been estimated at 20% of total production. The Food and Agricultural Organization (FAO) estimates losses of such crops on a worldwide basis at 10%. Individual crops may show much higher figures (e.g., sweet potatoes up to 95%), especially in the tropics, where it is estimated that, overall, 25% of all perishable food crops harvested are lost before consumption.

The storage of food is a technical field with little or no glamour associated with it, but there is great scope for real improvement in the world's food supply if more attention and resources were given to this vital area.

In addition to losses caused by organisms eating, fouling, or decaying stored foods, toxins may also be liberated by bacteria and fungi. The consumption of mouldy grain by humans and animals may have serious consequences if fungal toxins (mycotoxins) are present. Mycotoxins are considered later in this chapter. Some examples of the problems encountered in comestible produce are given in Tables 2.1 and 2.2.

Foods are protected from decay in a variety of ways, many of them traditional and of long standing. Drying reduces the water activity (see Chapter 6), as does the addition of sugars and salt, which also give high osmotic conditions in solution, making microbial attack less likely. Changing the environment to one of high acidity by pickling in vinegar is a technique which is centuries old. The addition of chemical preservatives is restricted to the use of compounds which appear harmless to humans (generally regarded as safe or GRAS), and the choice of such preservatives is rather limited. In bulk store, physical methods have their advantages: Refrigeration and inert gas atmospheres are employed extensively. Deep-frozen foods are now common in domestic storage worldwide. Modified gas atmospheres are increasingly used for retail packaging of food. Low oxygen (2%) limits microbial growth and chemical oxidation, and packaging in special plastic film which scavenges naturally produced ethylene slows ripening and discolouration of vegetables. Heat sterilization, as performed in canning, can be extremely effective when carried out properly, although the taste and texture of some canned foods may be different from fresh-cooked foods. High-pressure treatment of some foods, as an alternative to heat sterilization, may be employed to maximize nutrient levels in products such as fruit juice. In recent years, the use of gamma irradiation has been shown to be effective, practical, and economic. First used for animal feeds, irradiation is now used in some

Table 2.1. Examples of fresh produce spoilage caused by microorganisms

Product	Examples of spoilage organisms	Characteristics of spoilage
Fresh meat	*Lactobacillus* spp., *Bacillus* spp., *Clostridium* spp., *Pseudomonas* spp.	Slime production on surface, putrefaction (breakdown of protein and lipids)
Cured meat	*Lactobacillus* spp., *Aspergillus* spp., *Penicillium* spp.	Greening and moulding
Fresh fruit and vegetables	*Erwinia carotovora, Botrytis cinerea*	Soft rots
Bread	*Penicillium* spp., *Rhizopus* spp., *Bacillus subtilis, Leuconostoc dextranum*	Moulding and ropiness (in dough or bread)
Tobacco processed with humectants	*Aspergillus niger, Aspergillus glaucus* group	Moulding in packets
Medium and low acid foods (pH 4.5 or more) e.g., meat, seafoods, vegetables	*Clostridium* spp., *Bacillus subtilis, Bacillus cereus*	Souring of food can produce swelling that is due to carbon dioxide production, flat sours
Acid products, fruit, and fruit juices	Yeasts, *Bacillus* spp. (*B. thermoacidurans*), *Lactobacillus* spp.	Fermentation, gas production, yeasty odour, flat sours, fermentation, acid and gas production

Table 2.2. Some major food-poisoning bacteria

Organism	Cause of poisoning
Escherichia coli strains	Infection
Campylobacter jejuni	Infection
Salmonella typhimurium	Infection
Staphylococcus aureus	Enterotoxin
Clostridium welchii	Enterotoxin
Clostridium botulinum	Enterotoxin
Bacillus cereus	Enterotoxin

Note: This list is in no way exhaustive, and effects can vary greatly from strain to strain.

situations for human foodstuffs, sometimes in conjunction with antimicrobial packaging.

Mycotoxins

In 1960, following the death of poultry in the UK subsequently linked to
the use of mouldy peanut meal from a common source in the feed, and the
death of trout on fish farms in the United States, a series of investigations
was initiated, leading to the discovery of a previously unrecognized range
of fungal toxins, the mycotoxins, which have great significance in human
and animal health.

The fact that some fungi are poisonous has been recognized since ancient times. Poisonous mushrooms and toadstools (loose terms but commonly used) feature in the folklore of many nations, but poisoning by
eating the fruit body of mushrooms and toadstools, termed mycetism,
differs from mycotoxicosis, in which a foodstuff contaminated with a mi-
crofungus and its toxic metabolites is consumed.

The reader is recommended to refer to the account of the discovery
of mycotoxins by Linsell (1977). The source of the first identified toxins was *Aspergillus flavus*, giving rise to the name aflatoxin. Since that
time, similar toxins have been found to be produced by other fungi, and
these toxins are grouped, with others, together under the broad term mycotoxins. Examples of mycotoxins other than aflatoxins include patulin,
produced by species of *Penicillium* and *Aspergillus* on apples and grain;
ochratoxin, also produced by *Aspergillus* and *Penicillium* species on grain;
the trichothecenes, a large group of mycotoxins, produced by species of
Fusarium, Mycothecium, and *Stachybotrys* on cereals, grass, hay and grain;
zearalenone produced by at least four species of *Fusarium* on grain (often
maize), and sporodesmin, produced by *Pithomyces chartarum* growing
on grass litter, which causes a disease in cattle and sheep identified by the
condition known as facial eczema.

Fungi which have the potential to produce mycotoxins are common
and found worldwide in soil and the atmosphere, but the ability to actually synthesize these compounds varies greatly with the precise fungal
strain, the growth conditions, and the type of substrate. Favourable conditions for mycotoxin production occur where humidity and moisture are
high; grains and nuts tend to be the most susceptible materials. Chemically, the aflatoxins are a group of bisfurano isocoumarin metabolites,
synthesized by *A. flavus* and *A. parasiticus*. Unfortunately, these toxins are

Table 2.3. Toxicity of aflatoxin in various animal species

Species	Acute LD_{50} (mg/kg body weight)
Rabbit	0.30–0.50
Duckling	0.35–0.56
Dog	1.00
Monkey	2.20
Rainbow trout	0.81
Hamster	10.20
Sheep	2.00

Source: Adapted from Linsell, 1977.

stable under normal food-handling and cooking conditions, particularly when adsorbed on starch or proteins in cereal grains and other seeds. The only process which appears to have a marked detoxifying effect is strong alkali treatment, but this is not common in food preparation.

Aflatoxins are highly toxic, but toxicity varies on ingestion by different animal species (see Table 2.3). Apart from being poisonous, these compounds have been shown to be linked to several forms of cancer. Aflatoxin B_1 is a potent carcinogen when administered to experimental animals, and there is a strong correlation between the incidence of human liver cancer and the consumption of foods containing this toxin in areas of South-East Asia and East and West Africa. The young and sick are mainly at risk, but those with existing liver problems are particularly susceptible. There may be a synergism between mycotoxins and the causative agents of viral infections such as viral hepatitis. In the developed countries of the world, mouldy food would be rejected by most consumers; in the poorer areas of the world the population often has little choice of food quality and problems such as those just discussed inevitably result. However, the significance of even very low levels of mycotoxins in foods has yet to be established.

INSECTS AND MITES IN STORED PRODUCTS

Apart from the microorganisms already mentioned and the rodents (see Chapter 4), insects are the other main group of organisms to attack stored products. Throughout the world there are many different insects which commonly attack stored foodstuffs. The subsequent examples relate particularly to the UK, but serve to illustrate typical problems. Farm-stored grain is considered separately from general food stores for, as will be seen, the effects of insect attack on bulk grain can lead to other problems.

Order Lepidoptera

The warehouse moth (*Ephestia elutella*)

This small, greyish moth is common in the UK. The adults are seen between May and September, but the damage is done by the larvae, which are about 12 mm long and cream in colour, leaving strands of webbing as they emerge from food stacks to find cracks and crevices in the building structure in which they pupate. Wide varieties of foods are attacked, including chocolate, raw cocoa beans, cereals and cereal products, dried fruits, spices, nuts, and tobacco. Infestation is maintained by the fact that part of the year the pupae are hidden in the building structure, and therefore the infestation persists despite the turnover of stock.

The mill moth or mediterranean flour moth (*E. kuhniella*)

This is similar to, but larger and darker than, the warehouse moth. The larvae do not migrate out from the food they infest, but over-winter within it as pupae. This moth mainly attacks flour and cereal products. Large amounts of webbing are often produced by the larvae, and this can choke grain transport machinery, spouts, and pipes. From flour mills, the insect has been known to travel in retail packets to the consumer.

The Indian meal moth (*Plodia interpunctella*)

The Indian meal moth is commonly found in nuts, dried fruit and pet foods. In recent years it has become established as a domestic pest in houses and shops in many countries in Northern Europe, including Southern England.

There are several other moths found in UK product stores, including the tropical warehouse moth (*E. cautella*) and the rice moth (*Corcyra cephalonica*). These moths originate in warmer climes and do best as pests in heated premises.

Order Coleoptera

The Australian spider beetle (*Ptinus tectus*)
These are small round beetles with a superficial resemblance to spiders. They are nocturnal, and heavy infestations can build up before detection. They are omnivorous and also damage sacks, paper wrappings, and cardboard. Birds' nests, animal droppings, and carrion may also harbour this insect, and the adults are able to damage textiles.

The biscuit beetle (*Stegobium paniceum*)
The biscuit beetle mainly attacks dry hard cereals and cereal products, in which the larvae make small holes similar to those made by woodworm larvae. This is the insect which plagued generations of old-time sailors by infesting their ship's biscuits. It can also cause considerable damage to sacking, paper, and card and has even been reported to pierce metal foils. The tiny larvae are very active and can crawl round non-glued folds of food packets to get into the contents.

The confused flour beetle (*Tribolium confusum*)
This beetle is also mainly found in cereal products and flour, but can also attack a range of other foods, including dried fruit, beans, and chocolate products. It is mainly a pest in bulk stores, but can also be found from time to time in domestic premises. Like some of the insects mentioned in the next section, relating to farm-stored grain, this beetle prefers a high temperature, around 25 °C, and will not breed or develop below 18 °C. This is also true of the related rust-red flour beetle (*T. castaneum*), which requires even higher temperatures.

Order Acarina

Food mites and other mites
These are very small invertebrates, related to spiders, and may appear to the observer as 'living dust'. Mites require high humidity [about 65% relative humidity (RH) or more] and flourish in badly ventilated conditions or in and around high-moisture foods. Common species are the flour mite (*Acarus siro*) and the prune or dried-fruit mite (*Carpoglyphus lactis*). This latter species can give rise to an eczema-like condition in people handling infested foodstuffs, known as 'grocer's itch'. In damp conditions in houses, the house or furniture mite (*Glycyphagus domesticus*) may be

found, browsing on mould, and also the dust mite (*Dermatophagoides pteronyssimus*) feeding on desquamated human skin scales (the source of most of the fine white dust found in bedrooms). There is a probable relationship between allergies to 'house dust' and dust or house mites and their droppings. Great improvements in such allergies (e.g., bronchial asthma) have been reported following a strict regime of vacuum cleaning seams in bedding and along edges of carpets, followed by daily airing of bedding to reduce the high-humidity conditions favoured by these organisms.

Order Psocoptera

Booklice (*Psocids*)

In recent years, more and more interest has been shown in these insects, especially in bulk storage areas and domestic households and in buildings where straw-based interior partitioning is used. There are a number of species, of which *Liposcelis bostrychophila* is the most common in food and damp paper. Booklice are small invertebrates which have the ability to penetrate folded seams of packets or any container not tightly and totally sealed. They prefer humid conditions and starchy foods such as flour and cereals. Reproduction is rapid, and heavy infestations can build up quickly. Booklice are not related to the true lice, and those found on books are merely browsers on mould resulting from damp storage.

PREVENTION OF INFESTATION BY PESTS OF STORED PRODUCTS

The importance of careful, thorough, and regular cleaning of premises as a normal measure cannot be overstressed. This rule applies not only to warehouses, but also to stores generally, whether of books and documents, as in archives and libraries, or of furniture as in the case of domestic premises.

Incoming goods should be examined on arrival, preferably in a separate receiving area. This is easy to say but often difficult or impossible to do in practice. Different consignments of goods should be kept separate and well stacked, well away from walls. There is a human tendency when a person is filling an empty room for that person to start by stacking materials against all the walls. This hides some of the surface of the packs from view, creates harbourage and a means of transfer of organisms between the structure and the goods, and may create a more suitable microclimate for microorganisms to flourish. This wall-stacking tendency should be discouraged wherever possible.

Stock should always be used in rotation and the atmosphere kept dry and cool. The building itself should be kept in good repair. With old buildings, or buildings not originally designed as warehouses, there are often attempts made to 'modernize' conditions. In some cases storage conditions can be made worse by such 'improvements', which often take the form of panelling-in shallow alcoves and pipework or covering painted brick with sheets of plywood or building board. (This practice is also common in hospitals.) The effect is to create dead spaces which are difficult to clean, treat, or inspect, unless the panelling is removed or access panels are provided. Debris can accumulate and infestations develop unseen and unchecked. There are usually plenty of gaps to allow pests free access to such areas. Great thought should be given to any envisaged alterations to storage areas.

Packaging of goods is important, and there are several general pointers worth considering. Good seals are needed, and thicker materials are usually better than flimsy thin ones. Foil laminates which are now widely used for some products are particularly good against ingress of insects. Smooth surfaces are generally better than rough or sculptured ones. Manufacturing plant design often ignores biological considerations, and inspection and cleaning schedules often have scope for improvement. A key component of pest prevention in food processing and storage is monitoring by use of traps. Cleaning and control measures can then be targeted on the infested zones. Very effective attractant pheromone lures are available for some species, such as the warehouse moths *Ephestia* sp. and Indian meal moth *Plodia*.

When good housekeeping proves insufficient to control insects, chemical control methods may have to be employed. There are two basic types, fumigation and the use of contact insecticides. Fumigation employs poisonous gases or volatile liquids and is a very specialized job, for reasons both of technique and safety. Contact insecticides are more commonly used; however, as the name implies, to be effective the chemicals must come into intimate contact with the insects. A layer of dust can protect insects, and those living deep within food may remain safe. The use of insecticides is no substitute for good housekeeping; a high standard of cleanliness is necessary for insecticides to be efficient when used.

Insecticides can be applied as sprays, dusts, or smokes. Sprays are most suitable for treatment of walls, floors, and pallets. Dusts are useful for treating dead spaces such as ducting and false panelling. Smokes should be used only for treating inaccessible voids and roof spaces. In the past, lindane (gamma hexachlorocyclohexane) was widely used, but this is now banned in many countries. Organophosphorus (OP) insecticides, such as malathion, fenitrothion, and pirimiphos methyl, are effective, but use of OPs is restricted in many food-producing plants. Synthetic pyrethroids, including permethrin, deltamethrin, and cypermethrin, are very effective and are now widely used in the food industry. In some food-production and food-handling areas the only pesticides which are allowed are natural pyrethrins. 2,2-dichlorovinyl dimethyl phosphate (DDVP; it is also known by its trade name, Dichlorvos) is banned in food plants in many countries, including the UK. Always check laws and regulations governing the use of pesticides, as these will vary in different parts of the world.

Infestations of stored-product insects in commodities can be controlled by fumigation under sheets with methyl bromide or phosphine. This should only be carried out by a contractor qualified to use fumigant gases. Methyl bromide will be phased out in many countries by 2004 and will eventually be withdrawn in the developing countries.

FARM-STORED (BULK-STORED) GRAIN

Cool, dry grain, not damp enough to support microorganisms and therefore not 'self-heating', can last for years if not attacked by insects. Deep dry tombs in arid areas of the world have yielded samples of grain, centuries old, in apparently good condition.

Grain in bulk storage can be infested with insects from several sources, a common example being bought-in animal feeds on farms, especially unprocessed cereals and oilseed cakes. Sacks may also be sources of contamination. In the UK there are about twelve insect species which attack grain. Some attack only damp grain, and their presence is an indication that the grain is in poor condition. Particularly important pests are the saw-toothed grain beetle (*Oryzaephilus surinamensis*), the grain weevil (*Sitophilus granarius*), and the flat grain beetle (*Cryptolestes ferrugineus*). *Oryzaephilus* and *Sitophilus* do best in grain over 14% moisture. *Oryzaephilus* will not breed below 18 °C, whereas *Sitophilus* will breed down to 13 °C, but multiplication is slow at this temperature. Thus in temperate zones, including the UK, temperature is the key factor.

When a large quantity (tonnes) of grain stored in bulk in a heap, bin, or silo is attacked by insects, usually well within the grain, the heat produced by their metabolism causes a rise in temperature, aided by the good insulatory properties of the grain. This in turn causes a migration of water, which condenses out on the cooler outer layers of grain. If there is sufficient condensed water, the grain will then germinate, causing caking of the grain. This in itself is a serious problem, but the caked grain also encourages mould growth, which accelerates the heating problem. The resulting caked and mouldy grain leads to greater problems and economic loss than caused by the insect attack alone.

Precautions to prevent such damage are necessary and include the provision of cool, sound, well-ventilated buildings free of cracks. Where attack is likely, grain can be treated with a prophylactic insecticide to prevent damage. Malathion has been widely used in the past, but serious resistance to this insecticide has developed in some species. Pirimiphos methyl and chlorpyriphos methyl are also widely used on grain, but, as all admixture applications leave residues of pesticide in grain and grain products, their use has been restricted in many countries.

Large bulks of infested grain can be fumigated with phosphine gas released from aluminium or magnesium phosphide. More recently, systems delivering gaseous phosphine have been developed. In some countries there has been increasing resistance to phosphine in some insect species, and carbon dioxide has been used as an alternative for treating grain. Regulations allowing the use of such chemicals vary in different parts of the world. After harvest, home-grown grain should be kept separate from old or bought-in grain. The crop should be screened to remove plant debris (trash) and dried. The grain should be cooled down to 18 °C or less. Grain bins are often ventilated by air which is blown from the bottom. In the UK this is effective in winter when the RH is below 80% and in summer when below 75%. Grain should be inspected at intervals and the temperature taken at depth. Temperature rises can then be dealt with by increased ventilation as a first step, and chemical measures taken at an early stage if an infestation becomes established. In the UK, the Department for the Environment, Food, the Regions and Rural Affairs (DEFRA) is able to supply expert advice and literature on all aspects of grain storage and treatment.

Biodeterioration of natural products of animal origin

LEATHER

When animals are flayed after slaughter, the main interest is usually in the meat. The skins, which will eventually be used to produce leather, rarely, if ever, have the same care of handling. Cattle hides usually remain in a pile for some hours, only slowly cooling from the body heat of the animal and contaminated with urine, faeces, and body fluids. In these first few hours there is intense activity by lipolytic and proteolytic bacteria resulting in degradation of the fibre structure within, particularly in the grain enamel. This is important as the grain enamel gives the leather its much-desired surface texture. Spraying with biocides can help, but it appears that cooling to around ambient temperature by simply hanging hides on racks and blowing air over them with electric fans is of great benefit at this stage. Later in leather processing, particularly at the wet-blue stage after chrome tanning, but while the hides are still wet, the fungi (species of *Aspergillus* and *Penicillium*) take over a greater role in leather deterioration. The production of some leathers may take many months, and while in their wet condition they may be susceptible to further attack. The moderate acidity and high moisture content of wet-blue hides encourages mould growth, resulting in staining and resistance to the dyeing process. Prolonged soaking at elevated temperatures encourages bacterial growth, which again affects the quality of the hide. A range of bacteria (*Bacillus megaterium*, *B. subtilis*, *B. pumilis*, and *Pseudomonas aeruginosa*) have been shown to cause perforations in goat and cattle hides after only 3- and 6-day soaking, respectively. Even where biocides have been used in the water, the extracellular proteases liberated by the bacteria are able to continue degrading the hides. After soaking, liming, and de-liming, the hides are tanned.

Chrome-tanned leather has been found to be affected by 'red spots', which are growths of fungi from the genera *Penicillium* and *Paecilomyces*. These fungi are tolerant to the chromium compounds used in this form of tanning and dominate under the drier conditions of the product. During the drying process after tanning, the humidity and temperature may encourage some fungi to colonize the leather. To discourage this, the temperature can be reduced or fungicides may be employed. Once finished, chrome-tanned leather is usually highly resistant to mould growth, although, under adverse conditions, at least superficial mould growth can occur. A classic example of biodeterioration, as once used on one laboratory's publicity material, is the spectacular surface mould growth exhibited by a pair of used leather boots which had been left unattended for 48 h in damp tropical conditions. Vegetable-tanned leather is one of the most easily attacked of all materials, and some very fine supple leathers, such as those used for expensive clothing, decorations, and bookbinding, may contain sugars when finished which increase the risk of disfigurement and attack in poor storage conditions. Semi-chrome leather is intermediate in its resistance, seldom supporting more than slight growth of mould, but never completely resistant.

Untanned leather and hides can also be attacked by insects, notably by the Dermestid beetles – the hide (skin) and larder (bacon) beetles. Tanned leather in good condition is rarely attacked. In nature, these beetles are scavengers on various forms of animal waste, and some species have been used to strip down delicate animal skeletons, leaving the bones to be made into museum exhibits. These insects can be found in tanneries, skin and hide warehouses, bone-processing plants, and pet food factories. Stuffed animals and other museum specimens of animal origin can also be attacked, infestations perhaps spreading from the nests or dead bodies of birds or rodents in and around the store or museum. In libraries, the larvae of Dermestid beetles can do damage, not only by eating leather bindings or parchment, but by burrowing deeply into books, making tunnels in which to pupate.

WOOL, FUR, FEATHERS, AND MUSEUM SPECIMENS

The Dermestids (and some other beetles) and a variety of moths are the main deteriogens of woollen textiles. As with fur and feathers, wool consists mainly of keratin, a highly insoluble protein containing sulfur linkages in the molecule. Some microorganisms are able to produce enzymes to break down keratin, including some species of streptomycetes and some fungi which act as dermatophytes of man and animals; an example is one of the causative organisms of ringworm, the fungus *Trichophyton*. Keratinophilic fungi are also present in the soil, where in nature most terrestrial animal remains are broken down. Such fungi are also commonly found on living birds and animals; *Trichophyton* has been isolated from the fur of moulting elephant seals in the Antarctic by one of the authors (DA).

Newly shorn sheep wool contains additional nutrients, in the form of lanolin and dirt, and microbial attack can occur if storage is poor. After the wool is washed and de-greased, the possibility of attack is reduced. A similar situation exists with another animal fibre, silk, which is more resistant to attack when completely de-gummed. The natural preening oils found on birds' feathers, however, may inhibit fungal growth. Not surprisingly, keratinophilic fungi are found in abundance in areas where the environment has been enriched with keratin. An outbreak of fungal infection of the fingernails and the skin on the hands of farm workers occurred when unsterilized waste (heads, beaks, feet) from a poultry-packing plant was used to fertilize soil, a most unsound practice.

Wool, fur, and feathers are usually to be found as clothing, furnishings, and decorative items. Normal domestic care of such materials usually precludes attack by microorganisms, unless they are kept in very poor storage conditions; the main deteriogens are usually insects.

Moths

These include the common clothes moth (*Tineola bisselliella*), the tapestry moth (*Trichophaga tapetzella*), the case-bearing clothes moth (*Tinea pellionella*), and the brown house moth (*Hofmannophila pseudospretella*). It is the larvae of these moths which damage woollen textiles, fur, and feathers, although some require warm conditions and nutrients other than just wool to thrive. Dirty clothes and textiles are more attractive, providing extra proteins, amino acids, and the vitamin B complex. Some species, such as *Hofmannophila*, thrive only in damp conditions. The larvae of most species live in the textiles or animal skins and produce tubes or sheets of webbing. The case-bearing clothes moth larva moves around freely in a silk case. In nature, many moths and beetles obtain their nutrients (especially keratin) from birds' nests, which, as previously mentioned, are commonly reservoirs for a variety of pests.

Dermestid beetles

As deteriogens, the most important genera within the family Dermestidae are *Anthrenus, Attagenus, Dermestes,* and *Trogoderma*. Most are general scavengers on dead-animal material [a notable exception being the Khapra beetle (*T. granarium*), a serious tropical and subtropical pest of grain]. Some examples of Dermestid beetles which are common pests of animal-derived materials are given in the subsequent subsection.

Carpet beetles (*Anthrenus* spp.)

The adults feed outdoors on pollen and nectar, and eggs are laid on suitable materials, the larvae feeding on feathers, hair, and other animal detritus. They are adapted to the low water content of such materials. The yellow-brown hairy larvae are called 'woolly-bears', but are quite unlike the caterpillars of some butterflies and moths which are sometimes given the same name. These larvae will attack wool (often clean new material), fur, including preserved animal specimens, and dead insects, the latter item of diet making them a feared pest of insect collections in museums. The fur beetle (*Attagenus pellio*) has a similar lifestyle to that of the carpet beetles.

It should be noted that common names for these beetles cross genera; there are various kinds of carpet beetle in both the *Attagenus* and *Anthrenus* genera.

Dermestes spp.

These have already been mentioned in connection with leather. Members of this genus are the largest Dermestids found in stored products. There are five common species, which are more or less cosmopolitan, including the larder or bacon beetle (*D. lardarius*), once a common pest of dried meat in homes and a general pest of materials of animal origin, and *D. maculatus*, a serious pest of furs, hides, and skins. The Peruvian hide beetle, *D. peruvianus*, is an increasingly common pest in catering and housing in urban areas. The larvae feed on protein debris.

Other pests

Woollen textiles and other materials of animal origin can be attacked by other organisms, including rodents collecting nesting material and also a range of insects which are omnivorous or do incidental damage while scavenging for other foods, including Ptinid beetles (spider beetles), house mites (feeding on fungi), silverfish, and cockroaches.

ANIMAL GLUE

For many applications, animal glue has been superseded by synthetic products (see Chapter 3) which are more convenient to use. This is particularly true of hard glues used in woodworking. Animal glues are still used, however, and many materials which were made using them in the past are still in existence. It is thought that the peak time of woodworm (*Anobium*) infestation in the UK coincided with the maximum use of plywood made with animal glue, an addition which makes the product much more nutritious to the *Anobium* larvae.

Animal glues are usually aqueous solutions of proteinaceous material. They may contain gelatin or casein, both of which when in the liquid state are highly susceptible to microbial attack. During factory production, infected products may give off ammoniacal or sulfurous odours, especially after standing overnight, and the glue itself will be adversely affected and show thinning. Lower levels of infection may show themselves in retail packs only after longer storage. Generally there is a need for the addition of a chemical preservative during manufacture to give in-can protection of such products. It is interesting to note that adhesives made of cellulose-derived compounds (usually carboxymethyl cellulose), such as wallpaper paste, sold as dried granules, often contain a fungicide to prevent mould growth after application, a common occurrence if walls are damp or if condensation occurs.

CONTROL OF DETERIORATION OF WOOL AND OTHER ANIMAL-DERIVED PRODUCTS

Some materials, such as woollen carpets, are often mothproofed by chemical treatment during manufacture. Over the years, there has been widespread but often overoptimistic use of mothballs made of naphthalene or paradichlorobenzene (PDB). These products need to be used in considerable amounts (1 kg/m^3 or more), and even then their effectiveness is more as a repellent than in killing an established infestation. Use of PDB can give rise to deleterious effects (in combination with heat and light) to colours and physical properties of some textiles, and there are some health risks to those people continually exposed to the vapour. The Ethnographic Museum of Stockholm, faced with removing PDB from large quantities of textiles being taken from long-term storage for eventual display, used a specially developed low-pressure ventilation apparatus which utilizes low-frequency sound and forced ventilation to remove PDB from fragile materials.

Fumigation with methyl bromide will kill insect larvae and eggs deep in furniture or textiles. This must be carried out in a chamber or under sheets by a qualified operator. Because of concerns over safety and pesticide residues, alternative physical treatments are now becoming more widely used. Low-temperature treatments of bagged textiles for 3 days at minus 30 °C or 2 weeks at minus 18 °C will kill pests. Furniture or textiles can also be treated by carbon dioxide fumigation or nitrogen anoxia in oxygen barrier film bags. Exposure generally needs to be of at least 3-week duration. A humidity-controlled heat chamber at 52 °C is a very quick and effective way of killing pests in furniture and textiles.

Application of residual insecticides directly to textiles and furniture is rarely effective. However, careful application of a pyrethroid to infested areas can achieve some control of infestation in fitted carpets.

The storage structures (drawer, cupboard, box) should also be treated. However, the safest, most common, and probably the most effective preventive measure is good housekeeping and cleaning. Brushing, beating, and vacuum cleaning of textiles and other goods destroys eggs, as does washing at 60 °C and dry cleaning. Many insect pests shun bright light and tend to prefer the conditions of long-term storage. Goods should be removed and inspected regularly if not in use.

Stone

Most types of microorganisms (bacteria, fungi, algae), lichens, and higher plants (see Chapter 4) have been reported to play various roles in the colonization and decay of natural stonework found in buildings and monuments and to play a similar role in the weathering of natural rock. 'Synthetic' stones such as concrete, brickwork and mortar also show colonization, once the initial highly alkaline pH (above 10) falls after a period of weathering. Microorganisms can colonize cement and renderings over a wide pH range (4–10).

MICROORGANISMS IMPLICATED IN STONE BIODETERIORATION

The extent to which microorganisms, particularly heterotrophic bacteria, are involved in the deterioration of stonework is still a matter of some argument. They are, at least, passive colonizers of stonework, leading to soiling, and various mechanisms have been suggested whereby they may act as more serious and active deteriogens of stone at some depth. Their effects are shown in Table 2.4

Algae and cyanobacteria

Algae are microscopic single-celled, filamentous, or colonial green plants which lack the vascular tissue and complex organs of higher plants, whereas cyanobacteria are photosynthetic bacteria, lacking the membrane-bound organelles (such as chloroplasts) of the algae; this

Table 2.4. Microbial activities which affect the durability of stone

Observed effect	Microbial activity	Major microorganisms involved
Discolouration	Presence of pigmented cells or products	Algae, cyanobacteria, fungi
Retention of water	Physical presence Production of slimes	All
Stimulation of growth of heterotrophic and higher organisms	Presence of live or dead cells or cell products	Algae, photosynthetic bacteria, including cyanobacteria
Disaggregation of material	Penetration into and growth within stone	Fungi, actinomycetes, cyanobacteria, algae (also lichens)
Formation of patinas	Oxidation of translocated cations	Iron and manganese oxidizing bacteria, fungi, cyanobacteria
Degradation ('corrosion')	Acid production	Fungi, bacteria, cyanobacteria (also lichens)
Weakening and dissolution of structure	Mobilization and chelation of ions	All
Alkaline dissolution	Uptake of H^+ ions by cells	Algae, cyanobacteria
Disruption of layered silicates	Liberation of polyols (e.g., glycerol, polysaccharides)	All

group was previously called blue-green algae. They can often be found growing on stonework where there is light and a source of water. Water may be present because of leaking pipes or faulty roof guttering, inadequate drainage of flat areas, frequent wetting by wind-blown rain, or from adjacent watercourses as in the case of stone walls and embankments along rivers and canals. In the humid tropics and sub-tropics, no obvious source of water is necessary, the moisture in the air being sufficient for growth.

These microorganisms may cause mechanical damage by colonizing and widening cracks in stone as a result of growth, especially when frost action enhances this effect. They may actually grow within the stone, as so-called endolithic organisms (Figure 2.2), thus directly causing disruption of the structure. They also produce organic acids, and instances have been reported in which calcium carbonate has been dissolved out of concrete by blue-green bacteria. Aesthetic biodeterioration or soiling is probably the most important type of damage caused by these microorganisms (apart from the fouling of ships' hulls by larger marine algae; see Chapter 4).

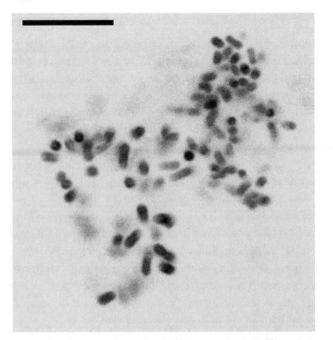

Figure 2.2. A microcolony of endolithic cyanobacteria (*Synechococcus*) growing within soapstone on a church in Minas Gerais, Brazil. Laser confocal scanning microscopy of autofluorescent cells in situ (bar length: 10 μm). Photo: Dr Christine Gaylarde.

This occurs not only on stone but on wood, painted surfaces, and, indeed, any surface exposed to a minimum level of sunlight. Large patches of green, pink, and brown growths give an unsightly appearance of neglect to buildings, and such growths can develop very quickly (a few weeks), especially in the humid tropics. The damper western areas of the UK show most algal soiling of stonework, often in rural areas, where wind-blown agricultural fertilizers can sometimes create very suitable nutrient conditions on damp stonework. The growth of dark algal and cyanobacterial slimes on flat paving and airport runways is not only unsightly, but can be slippery and dangerous. It also leads to the formation of a humus layer, encouraging growth of higher plants. This often happens on flat roofs which are poorly maintained. Some types of rough external renderings can assist colonization by providing many small sites protected from desiccation. However, the cyanobacteria are very resistant to desiccation and survive well on dry stone surfaces, even at high altitudes in the tropics, where they are protected from high UV levels by pigments within the cells.

Fungi

Fungi can also be found on stonework and, similarly, have the ability to withstand dry conditions and sometimes high UV levels. Citric and oxalic acid production by genera such as *Botrytis, Mucor, Penicillium*, and *Trichoderma* can result in the solubilization of silicates and the weathering of stone as a consequence. *A. flavus* has been shown to solubilize rock phosphate, at least in the laboratory. The chelation of iron by *Penicillium* and *Acremonium* (*Cephalosporium*) has also been demonstrated under laboratory conditions, and it has been postulated that this may occur in stone. Dark-pigmented mitosporic fungi have been shown to produce pits in limestone, and they are, of course, very important in the biofilms that cause surface discolorations. Fungi have also been implicated in concrete corrosion. A strain of *A. glaucus*, isolated from rotten concrete in a cellar in the UK, released high levels of calcium when incubated with sterile concrete blocks in a mineral salts–glucose medium. No released aluminium was detected, however, and it was suggested that this element was either chelated by organic acids and not detected by the system used, or was locked within the alumina silicate matrix of the concrete.

It has been suggested that fungal (and bacterial) deterioration of stone and concrete has increased since the Industrial Revolution. In urban and industrialized regions, a layer of carbonaceous pollution forms on buildings, and this may contain food material for heterotrophic microorganisms. The surface of the Seville cathedral, for example, is covered with a black crust which contains hydrocarbons. These can be used by certain fungi and bacteria, which produce acids as a result of their metabolism, hence corroding the underlying surface.

Bacteria

The part played by heterotrophic bacteria in the biodeterioration of stone is not at all clear, but the evidence that they do contribute to the breakdown of stone is increasing. One mechanism that has been suggested involves the production of calcium sulfate in stone. It is proposed that the sulfate-reducing bacteria, present in the soil, produce hydrogen sulfide as a result of their normal (anaerobic) metabolism and that this is then carried in solution up suitably porous stone walls by capillary action (as in rising damp). The autotrophic sulfur-oxidizing bacteria (which have been isolated from stone by several workers) are then able to oxidize the sulfide, utilizing it as an energy source, the result being the production of sulfuric acid. This acid then attacks the stone, producing calcium sulfate, which forms surface scales. Recent workers in Germany, however, were unable to detect any sulfuric acid-forming bacteria on natural stone surfaces, casting the mechanism into doubt for this substrate.

'Artificial' stone, in the form of concrete, has, on the other hand, been known since the 1940s to be subject to attack by biogenic sulfuric acid. New concrete has a pH of 12–13, and this must be reduced by reaction with atmospheric carbon dioxide (a process known as carbonation) before bacterial activity can begin. In the case of building surfaces, this is an extended process, but in concrete pipelines which carry fluids, as in sewerage systems, carbonation can occur much more rapidly and bacterial biodeterioration will be more important. In sewer pipes, sulfate-reducing bacteria may also be implicated. In the Middle East, where the flow rates are slow, sewage may remain in pipes for some time, at elevated temperatures, allowing fluctuating anaerobic and aerobic conditions, leading to the production of sulfuric acid and corrosion of the concrete.

The nitrifying bacteria *Nitrosomonas* and *Nitrobacter* have been suggested as agents responsible for another possible mechanism. These organisms have been found in deteriorated stone where sulfate levels were not excessive. It is suggested that they oxidize ammonia in the air, forming nitrates which then react with the relatively insoluble calcium carbonate of calcareous rock, forming the more soluble calcium nitrate. This is then leached out of the stone by rain, leaving a loose powder of silica particles. However, recent work suggests that this is not an important mechanism of stone or concrete decay.

Heterotrophic bacteria capable of dissolving silicates in sandstone pebbles have been isolated from soil above sandstone in railway tunnels. Heterotrophic bacteria also produce organic acids such as 2-ketogluconic, causing the release of insoluble phosphates and silicates from rocks. There is, however, little published work on the role of these bacteria in concrete corrosion.

Much of the work to date has centred on laboratory studies, and, until more detailed studies are carried out in the field, the role of bacteria in the breakdown of stone will remain unquantified and the subject of continued speculation and debate.

Lichens (also called lichenized fungi)

The lichens show some clear deteriogenic abilities. Lichens are a complex association of a fungus (mycobiont) and an alga or cyanobacterium (photobionts) acting as a single organism. Together, the microorganisms produce a simple, macroscopic structure known as a thallus. Thalli may take various forms, depending on the symbiotic partners. Crustose lichens adhere very closely to the stone and are difficult to remove. Foliose thalli penetrate less deeply, using only thread-like anchorage devices, and fruticose lichens are generally anchored to the substrate only by a button-like structure; these two types are relatively readily removed without causing much damage to the stone. The fourth type of lichen, occurring only on calcareous stones, is endolithic, growing within the substrate and often only seen when the fruiting bodies emerge from the stone. Lichens are found on stonework under a wide range of temperatures and humidity, from polar regions to the tropics. They are very hardy and are able to withstand prolonged periods of desiccation, reabsorbing water and swelling quickly once it becomes available again. Different lichens prefer either calcareous or siliceous rocks; some remain on the surface, and, in others, the mycobiont hyphae are able to penetrate the rock. These characteristics can result in the deterioration of stone, with particles flaking off for the following reasons:

1. expansion of the hyphae embedded in stone
2. successive expansion and contraction of the lichen colony on wetting and drying. The moisture contents of lichens may vary between 300% and 15% (as a percentage of dry weight) in 2–3 h
3. the trapping of water in the stone around the lichen, which leads to frost damage in cold climates, and the increased absorption of atmospheric acids (a major physicochemical cause of stone erosion) into the stone.

Lichens also excrete organic acids (mainly oxalic and "lichenic" acids) as metabolic wastes, and these compounds are able to chelate cationic components (calcium, silicon), removing them from the stone structure. Increasing permeability which is due to penetration of rhizines (the root structures of the lichen) allows ingress of water, causing further erosion of the stone. It has been suggested that crustose lichens may protect stone surfaces from non-biological weathering and that they should not be removed from the substrate without careful consideration. It is true that the removal (physical or chemical) of lichens may result in joint removal of the stone surface, leaving the resultant surface more porous and more susceptible to chemical and biological deterioration. Nevertheless, leaving the lichen in place certainly allows the normal deteriorative mechanisms to continue and is potentially dangerous.

The dirty appearance of many buildings (often in the past attributed to atmospheric dirt) is commonly due to the presence of microbial biofilms or of very small but widespread colonies of dark lichens, common even in large towns and cities, despite the sensitivity of most lichen species to atmospheric and industrial air pollution.

One particular problem of the disfigurement of stone by lichens is headstones on war graves. The Commonwealth War Graves Commission, which has a worldwide responsibility for war cemeteries, has been concerned with this problem, as the headstones must appear clean and neat at all times, without signs of neglect and with all inscriptions clear and legible. This concern has resulted in remedial treatment being carried out on a large scale, initially by physical methods. More recently, trials have been made with biocidal washes to good effect. A large number of biocides have been used, including quaternary ammonium salts, and herbicides, such as Diuron and dithiocarbamates. It is important that the product chosen should have a slightly alkaline or neutral pH and must not favour the formation of salts.

Massive growth of lichen, as may occur in the tropics, can with time obscure carvings and inscriptions on monuments, and the niches provided by such growth can lead to the accumulation of wind-blown detritus, nutrients, and water, encouraging the establishment of moss and plant growth. Colonization of stone by some lichens may be encouraged by the presence of bird droppings, which provide extra nutrients.

INVERTEBRATE DETERIOGENS OF STONE

Masonry bees

Several species of solitary bees, whose natural home is a burrow in soil or soft stone, are able to colonize soft mortar between brickwork. An example is Davies's Colletes (*Colletes daviesanus*). Where conditions are particularly suitable, considerable numbers of these insects may construct individual burrows with a resulting weakening of the brickwork. Insecticides can be applied, but the best answer is repointing with hard mortar. Such insects have caused some legal controversy in the UK. Householders have claimed on property insurance after such damage, only to find that, in legal terms, insects are not regarded as animals! Most insurance policies exclude wood-boring insect damage and the mention of cover against animals is usually confined to claims arising from physical damage caused by collisions of livestock such as horses and cattle.

REFERENCES AND SUGGESTED READING

Arai, H. (1993). Relationship between fungi and brown spots found in various materials. In *Biodeterioration of Cultural Property; Proceedings of the 2nd International Conference, Yokohama, 5–8 October*, pp. 320–36.

Ascaso, C., Wierzchos, J., and Castello, R. (1998). Study of the biogenic weathering of calcareous litharenite stones caused by lichen and endolithic microorganisms, *Int. Biodeterior. Biodeg.*, **42**, 29–38.

Carter, T. P., Best, D. J., and Seal, K. J. (1988). Studies on the colonisation and degradation of human hair by *Streptomyces*. In *Biodeterioration 7*. Houghton, D. R., Smith, R. N., and Eggins H. O. W. (Eds.). Elsevier Applied Science, New York, pp. 171–9.

Gaylarde, C. C., and Morton, L. H. G. (2002). Biodegradation of mineral materials. In *Encyclopaedia of Environmental Microbiology*, Bitton, G. (Ed.). Wiley, New York, pp. 516–27.

Gorbushina, A. A., and Krumbein, W. E. (2000). Rock-dwelling fungal communities: Diversity of life-styles and colony structure. In *Journey to Diverse Microbial Worlds*, Seckbach, J. (Ed.). Kluwer Academic, Boston/Dordrecht, The Netherlands, pp. 317–34.

Gorham, J. R. (Ed) (1987). *Insect and Mite Pests in Food*. USDA Handbook No. 655. U.S. Department of Agriculture, Beltsville, MD.

Grant, C. (1982). Fouling of terrestrial substrates by algae and implications for control. *Int. Biodeterior., Bull.*, **18**, 57–65.

Griffin, P. S., Indictor, N., and Koestler, R. J. (1991). The biodeterioration of stone: a review of deterioration mechanisms, conservation case histories and treatment. *Int. Biodeterior.*, **28**, 187–207.

Hawks, C., and Rowe, W. F. (1988). Deterioration of hair by airbourne microorganisms; implications for museum biological collections. In *Biodeterioration 7*: Houghton, D. R., Smith, R. N., and Eggins H. O. W. (Eds.). Elsevier Applied Science, New York, pp. 461–5.

Hoffmann, L. (1989). Algae of terrestrial habitats. *Bot. Rev.*, **55**, 77–105.

Kirk, T. K., and Shimada, M. (1985). Lignin biodegradation: The organisms involved and the physiology and biochemistry of degradation by white-rot fungi. In *Biosynthesis and Biodegradation of Wood Components*, Higuchi, T. (Ed.). Academic, San Diego, CA, pp. 579–605.

Linsell, C. A. (1977). Aflatoxins. In *Environment and Man, Vol. 6: The Chemical Environment*, Lenihan, J., and Fletcher, W. W. (Eds.). Blackie, Glasgow/London, pp. 121–36.

Lisci, M., Monte, M., and Pacini, E. (2003). Lichens and higher plants on stone: A review. *Int. Biodeterior. Biodeg.*, **51**, 1–17.

May, E., Lewis, F. J., Pereira, S., Tayler, S., Seaward, M. R. D., and Allsopp, D. (1993). Microbial deterioration of building stone – A review. *Biodeterior. Abstr.*, **7**, 109–23.

Montegut, D., Indictor, N., and Koestler, R. J. (1991). Fungal deterioration of cellulosic textiles: A review. *Int. Biodeterior. Bull.*, **28**, 209–26.

Mourier, H., Winding, O., and Sunesen, E. (1977). *Collins Guide to Wildlife in House and Home.* Collins, London.

Ortega-Calvo, J. J., Hernandez-Marine, H., and Saiz-Jimenez, C. (1991). Biodeterioration of building materials by cyanobacteria and algae. *Int. Biodeterior.* **28**, 165–86.

Petersen, K., Kuroczkin, J., Strzelczyk, A. B., and Krumbein, W. E. (1988). Distribution and effects of fungi on and in sandstone. In *Biodeterioration* 7 Houghton, D. R., Smith R. N., and Eggins H. O. W. (Eds.). Elsevier Applied Science, New York, Vol. 7, pp. 123–8.

Purvis, W. (2000). *Lichens.* The Natural History Museum, London. Life Series.

Samson, R. A., and van Reenen-Hoekstra, E. S. (1988). *Introduction to Food-Borne Fungi.* Centraalbureau voor Schimmelcultures, Baarn, The Netherlands.

Samson, R. A., Hoeckstra, E. S., Frisvad, J. C., and Filtenborg, O., (Eds) (2001). *Introduction to Food- and Airborne Fungi*, 6th ed. Centraalbureau voor Schimmelcultures, Utrecht, The Netherlands. 90-70351-42-0.

Smith, J. E., and Moss, M. O. (1985). *Mycotoxins; Formation, analysis and significance.* John Wiley & Sons Ltd., Chichester, UK.

Sterflinger, K., and Krumbein, W. E. (1997). Dematiaceous fungi as a major agent for biopitting on Mediterranean marbles and limestones, *Geomicrobiol. J.*, **14**, 219–25.

Strzelczyk, A. B. (1981). Stone. Microbial biodeterioration. In *Economic Microbiology*, Rose, A. H. (Ed.). Academic Press, London, Vol. 6, pp. 61–79.

Suto, M., and Tomita, F. (2001). Induction and metabolite repression mechanisms of cellulase in fungi. *J. Biosci. Bioeng.*, **92**, 305–11.

Thirumale, S., Rani, D. S., and Nand, K. (2001). Control of cellulase formation by trehalose in *Clostridium papyrosolvens* CFR-703. *Process Biochem.*, **37**, 241–5.

Watanabe, H., and Tokuda, G. (2001). Animal cellulases. *Cell. Mol. Life Sci.*, **58**, 1167–78.

Whiston, R. A., and Seal, K. J. (1988). The occurrence of cellulases in the earthworm *Eisenia foetida. Biol. Waste*, **25**, 234–42.

Winkler, E. M. (1976). Decay of building stones. In: *The conservation of stone II. Proceedings of the second International Symposium*, Rossi-Manaresi, R. (Ed.). Centro per la Conservazione della Sculture all'Aperto, Bologna, Italy, pp. 27–36.

Zanardini, E., Abbruscato, P., Ghedini, N., Realini, M., and Sorlini, C. (2000). Influence of atmospheric pollutants on the biodeterioration of stone. *Int. Biodeterior. Biodeg.*, **45**, 35–42.

3

Biodeterioration of Refined and Processed Materials

Biodeterioration of fuels and lubricants

Fuels and lubricants are derived for the most part from naturally occurring petroleum deposits and are directly utilized by a range of oxygen-requiring microorganisms (upwards of 1000 species). Interest has recently been focused on the fate of waste oil in the environment, and a number of oil-degrading strains of bacteria have been isolated which may have biotechnological uses. Being essentially hydrophobic materials, fuels and lubricants are at risk of microbial degradation only when they are in close contact with water. The water may be present as a discrete layer beneath the hydrocarbon, as large dispersed droplets in an agitated system or as an oil-in-water or water-in-oil emulsion, where the water droplet sizes are in the range of 1–10 μm in diameter. The presence of only 10 parts per million (ppm) of water may be sufficient for bacteria or fungi to grow at the interface between the oil, or fuel, and the water, and fuels containing as much as 0.1% water will normally meet the quality standards for use. Hence even apparently water-free hydrocarbons may suffer from microbial contamination.

Some microorganisms excrete surface-active materials (biosurfactants), which allow the cells to contact the non-polar hydrocarbon molecules before the uptake and/or degradation of the latter. Figure 3.1 shows the emulsion produced between an aqueous medium and diesel oil by supernatants from cultures of *Pseudomonas* isolates, demonstrating that the emulsifying agents are extracellular.

Extracellular surfactants can cause significant problems in fuel storage tanks, causing emulsion (fuel-in-water) and invert emulsion (water-in-fuel) formation with the water bottom. The latter is especially important because polar molecules become dispersed in the fuel, reducing stability,

Figure 3.1. Emulsion between diesel oil and aqueous medium in which isolates of *Pseudomonas* have been cultured. T, Control medium without inoculation; isolates 25 and 29 have produced almost complete, stable emulsification of the oil. Photo: Helenice Silva de Jesus.

accelerating sedimentation, and, in the case of organic acids, increasing the fuel acid number and its corrosivity.

Fuels and lubricants are used in a wide variety of applications which will be familiar to the reader. Those applications in which water is not generally present or in which the temperature is, on average, too high for microbial growth (such as car engines) need not concern us here. The fractions which contain high levels of hydrocarbons (alkanes) with a chain length of between ten and eighteen carbon (C10–18) atoms are most likely to be utilized by microorganisms as a carbon and an energy source (see Table 3.1). As a result, fractions such as automobile gasoline (C5–12) do not suffer significant problems, whereas diesel, kerosene, and light oils (mineral and vegetable) have been highlighted in the literature and subsequently studied. At the other end of the spectrum, aromatic hydrocarbons have been found to be more resistant, probably as a result of their phenolic

Table 3.1. Fuel fractions obtained
from crude oil

Fraction	C atoms
Gas	1–4
Gasoline	5–12
Kerosene	10–16
Diesel	15–22

content, although metabolic pathways have been determined to account for their ultimate degradation in the environment. Such hydrocarbons are used less and less now because of irritant and carcinogenic effects. As will be subsequently seen, the presence of other components in the finished product can retard or enhance microbial growth.

A number of deterioration mechanisms have been identified which are common to all the applications of fuels and lubricants:

1. utilization of the hydrocarbon components resulting in build-up of biomass in the system,
2. degradation of the additives resulting in both biomass production and the loss of one or more performance properties, such as corrosion protection,
3. the formation of metabolic products such as hydrogen sulfide and organic acids, which create a corrosive environment to metals (see the section on metals).

The main areas of concern are hydrocarbon fuels in aircraft, ships, and road vehicles (particularly in storage tanks) and emulsions in the metal-working industry. Occasional problems occur with lubricants alone and also with hydraulic systems which can become contaminated with water. The rules which govern these latter situations and the mechanisms which occur are essentially the same as those previously outlined.

FUELS

Crude oil is a mixture of straight, branched, and cyclic aliphatics and aromatic and heterocyclic compounds. The refinery products are also mixtures; gasoline, for example, containing straight-chain and branched-chain hydrocarbons, alkenes, naphthalenes, aromatics, and

Table 3.2. Consequences of microbial growth in fuel systems

Principal microorganisms involved	Consequence(s) of growth
Fungi and bacteria	Sludge formation
Fungi and bacteria	Increased suspended solids
Fungi and bacteria	Increased water content
Fungi and bacteria	Reduced filter life
Fungi and aerobic bacteria	Degradation of hydrocarbons
Fungi and aerobic bacteria	Fouling of injectors
Fungi and aerobic bacteria	Emulsification of oil and water by surfactant production
Fungi and anaerobic bacteria	Corrosion of storage tanks and metal pipework
Fungi and polymer-producing bacteria	Biofilm formation, leading to blockages of filters, pipes, and valves and incorrect readings from fuel probes
Fungi and bacteria	Breakdown of tank linings
Sulfate-reducing bacteria	Increased sulfur content of fuel, corrosion of tanks
Endotoxin-producing bacteria, sulfate-reducing bacteria, opportunistic pathogens	Health problems

other compounds. Other chemicals may be added to stabilize the fuel or enhance its properties, to inhibit corrosion, or to prevent icing. These additives may include aromatic amines and phenols, chelating agents, tetraethyl or tetramethyl lead, alcohols, and surfactants. Some of these may inhibit microbial growth, whereas others may stimulate it. Microorganisms may grow in any fuel where a minimum of water is present. The problems which may result from their growth are shown in Table 3.2.

Fuels may be classified as slightly or highly contaminated by the number of microorganisms present in the water bottoms: 10^5 bacteria/ml and 10^3–10^4 fungi/ml for slight contamination and 10^6–10^8 bacteria/ml and 10^4–10^6 fungi/ml for high contamination. A clean fuel will contain less than 50 organisms, whereas the associated water may carry bacteria in excess of 10^4 organisms/ml. Sulfate-reducing bacteria (SRB) are detected in only the water phase and only when the fuel is highly contaminated. There are differences among the degree of biodeterioration in different classes of fuel, caused by the chemical differences among the hydrocarbons present, the additives used, and the importance attached to maintaining specifications.

Gasolines and aviation fuels (the lighter petroleum fractions)

Reports in the 1930s through the early 1950s indicated that gasoline was subject to biodeterioration during storage in tanks with water in the bottom. The problems manifested themselves in the formation of a microbial sludge and the contamination of the fuel by hydrogen sulfide. The hydrogen sulfide, formed as a result of SRB activity, induced corrosion of metal pipelines and aircraft fuel pumps. It was suggested that prevention of the oxidation of the gasoline by antioxidants, such as naphthols, catechol, cresols, and amino–phenols, would have an antimicrobial effect. It was also thought, rather naively, that the inclusion of lecithin, to eliminate deposits in the carburettor and to prevent rusting and icing, would result in its preferential degradation and spare the direct utilization of the gasoline components. It has been demonstrated that bacterial contamination is probably enhanced when lecithin is present. Its ability to reduce interfacial tension and to concentrate at the gasoline/water interface has been considered to account for this, but obviously its use as a bacterial nutrient is an important factor. (Lecithin is also a powerful biocide neutralizer!) Alternative anti-icing compounds, generally alcohols or surfactants, previously considered to have mildly antimicrobial properties, can also provide food for some microorganisms. The oxygenate methyl tributyl ether (MTBE) has been shown to be rapidly removed from gasoline stored over contaminated water bottoms. It has been suggested that microbial growth occurs in gasoline only at the expense of such additives.

With the advent of the jet engine and the use of kerosene, attention switched to the problems caused by this readily utilizable hydrocarbon mixture. Kerosene will, in the presence of water (and other nutrients, invariably present in such systems), support the growth of a range of bacteria and fungi, some of which are able to utilize the fuel directly and others which are opportunistic colonizers utilizing contaminating nutrients (see Table 3.3). Those able to utilize the fuel directly produce a range of metabolic products (organic acids, proteins, alcohols, and growth factors), which can then be used by the non-utilizers. Additives, such as 2-methoxyethanol (2-ME), an anti-icing agent, can also be utilized by some microorganisms. Oxygen is able to diffuse quite easily through the deep layers of liquid hydrocarbon fuel to permit growth at the kerosene/water

Table 3.3. Some bacterial and fungal genera capable of growth in or isolated from fuels

Bacteria	Fungi	
Acinetobacter*	Acremonium*	Humicola
Alcaligenes*	Alternaria	Paecilomyces*
Bacillus*	Aspergillus*	Penicillium*
Clostridium	Aureobasidium	Phialophora
Corynebacterium	Candida*	Rhinocladiella*
Flavobacterium	Chaetomium	Rhodotorula*
Ochrobactrum	Cladosporium*	Trichoderma*
Pseudomonas*	Fusarium*	Trichosporon*
Serratia	Helminthosporium	Tritirachium
SRB*	Hormoconis*	Ulocladium

*Asterisks indicate those in which some species have been shown to grow in fuel.

interface. The presence of water in the flight tanks is responsible for the numerous microbiological problems which have occurred in both civil and military aircraft. The water enters the fuel at a variety of points in its refining and distribution network, often with other nutrients and microorganisms. The kerosene itself will have water dissolved in it, the amount depending on the temperature of the fuel (the solubility of water in kerosene is approximately equal in parts per million to the temperature of the fuel in degrees Fahrenheit). As the fuel cools, water will be released from solution to form a discrete layer. The subsequent growth of microorganisms will yield more water as a result of their metabolism.

One particular fungus, *Hormoconis resinae* (previously known as *Cladosporium resinae*; perfect state, *Amorphotheca resinae*), has been consistently isolated from contaminated kerosene in jet fuel systems. *H. resinae* is a soil-inhabiting fungus which can be selectively isolated from creosote-impregnated matchsticks placed in the soil. Because of its ability to tolerate creosote and grow in the presence of kerosene, much work has been carried out on the ecology and physiology of this fungus. It is able to metabolize hydrocarbons of chain length C10–C19, including *n*-alkanes, branched alkanes, and cyclic and aromatic rings. Its mycelial habit allows it to grow over surfaces where water is present and at the fuel/water interface, where it can become associated with the fuel by the formation of an emulsified layer. The layer so formed has an affinity for the fuel which is due to the production of surface-active compounds, and the mycelial pellicle can then be carried into the fuel distribution system of the aircraft with disastrous results. Its attachment to a metal surface within an integral aircraft fuel tank can result in the formation of differential aeration (microbial deposit) corrosion cells (see section on metals) which cause localized pitting and eventual perforation of the structure. The use of linings to prevent this brings its own problems; *H. resinae* is able to penetrate liners made from either butadiene–acrylonitrile or polyurethane formulations. Polysulfide coatings used to line concrete storage tanks have been found to be degraded by mixed populations of bacteria. Similar observations have been made with the use of polyester polyurethane foam baffles, where the fungi produce extensive matting and will invariably contribute to the overall infection of the fuel. The nutrient requirements of *H. resinae* are not particularly complex. Nitrogen and phosphorus appear to be the most important limiting nutrients, and their affinity for phosphorus gives them an advantage in a phosphorus-limited environment, such as that to be found in fuel systems. The low in-flight temperatures may also be advantageous to *H. resinae*, as spores of this species can survive repeated freeze–thaw cycles with a minimum temperature of $-22\,°C$. It has been found that *H. resinae* can survive in kerosene which is free of water and will sporulate freely in fuel/water mixtures. *Pseudomonas*, by contrast, is severely reduced in activity after only 1 day (20% survival) in fuel–water mixtures and can survive only a few hours in fuel in the absence of water.

The problems previously described will normally be initiated on the ground as fuel temperatures in flight are too low (down to $-40°C$) for growth to occur and any water in the fuel tanks will be in the form of ice. The grounding of aircraft in warm humid environments, where the fuel may reach $35\,°–40\,°C$, will encourage growth of the fungi in the tanks, leading, in the most severe conditions, to blockages of fuel lines, filters, and fuel pumps, and even causing malfunctioning of fuel gauges by biofilm formation. Aircraft spending less time on the ground are less susceptible, providing normal checking procedures are followed. However, if in such cases the ice has not melted in the tanks, then water will not be detected when the inspection taps are opened. Military aircraft and executive jets have been more prone to problems because of their intermittent use. The situation in supersonic aircraft is reversed. Fuel in the integral tanks situated at the extremities of the wings acts as a heat sink for heat from friction and that generated by the auxiliary system, resulting in temperatures of

90–100 °C, which are sufficient for self-sterilization. The outboard tanks normally empty quickly, whereas those inboard rarely empty, being used for trim purposes. As the fuel is moved around to assist trim, the inboard tanks will be at 40–55 °C and well aerated. It has been shown in a simulation of these conditions that *H. resinae* is a poor competitor compared to thermophilic fungi (those fungi with a temperature range for growth of between 20 and 55 °C). An albino strain of *Aspergillus fumigatus* (a thermotolerant species with a growth range of between 10 and at least 50 °C) has been observed as a dominant colonizer in simulation experiments. Temperatures of 65 °C and above have been found to be sufficient to completely inactivate *H. resinae* in fuels, which may account for its inability to become dominant in any fungal community developing in supersonic aircraft. Fortunately, because of rigorous checking procedures, no serious problems caused by fungi have occurred in practice in supersonic jets.

Diesel fuel

Land vehicle and marine diesel fuels are composed of a much wider range of hydrocarbons than aviation or domestic kerosene and contain longer carbon chain molecules. Less strict quality control than is the case with aviation fuel, together with the increased use of additives, has led to these becoming the major problem areas with respect to fuel biodeterioration, which has been reported in regions from Antarctica, through the equator, to North Alaska. At the poles microbial growth is slower, but, over prolonged periods without treatment, the level of biodeterioration is the same.

The problems with marine diesel are particularly acute in seawater-displaced systems, such as those used in some warships. Seawater contains a range of extraneous nutrients derived from the organic matter in the water, as well as microbial contamination. The type and concentration of nutrients will determine the colonizing flora and its extent. Fuel systems in marine diesels where water displacement is used contain filters, water separators and coalescers to remove solid matter and water before delivery to the engines. The presence of excess sludge of microbial origin, or biofilms on filter elements, can result in premature failure of these components and a failure to deliver the fuel, a highly undesirable situation in a warship which must be able to respond at a moment's notice. The need to consider biofilm organisms is highlighted by the example quoted in Hill and Hill (1993), in which seizure of all pumps and injectors on a small cargo

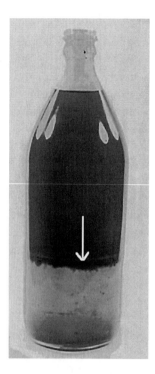

Figure 3.2. Fuel/water sample removed from a diesel storage tank in Brazil. Note heavy growth at the interface (indicated by arrow). Photo: Dr Fatima Bento.

ship was treated by physical cleaning, but no biocide addition. Hence live microorganisms remained attached to the walls of the system, with the result that the same problem recurred 2 months later when the vessel was at sea!

Land vehicle diesel fuels mainly show contamination problems in storage tanks, both in gas stations and at the refinery/distributor. A recent survey carried out in Brazil showed that the level of microbial contamination in diesel storage tanks at bus depots varied widely, with some foremen being unaware of the need to remove water from the tanks at regular intervals and hence promoting the formation of a water phase of low pH (down to 3) and considerable interfacial sludge (actually a liquid/liquid biofilm; see Figure 3.2). Highly contaminated fuels were associated with increased wear of engine parts and more frequent changing of filters.

Other light fuel oils such as central heating fuels and paraffin also support microbial growth. The authors have isolated *H. resinae, A. fumigatus, A. niger,* and *Paecilomyces variotii* from domestic paraffin. The results of the infections are often reported as equipment malfunctions.

CONTROL OF MICROBIAL GROWTH IN FUEL

Microbial contamination of fuel systems is kept to a minimum by attention to operational factors. The entry of airborne contaminants can be reduced by correct design of ventilation systems: the accumulation and ease of drainage of water are functions of the design of the whole system. In general terms, the dryer the system, the less microbial growth will occur. Biofilms can generally be expected to form on the walls of storage tanks, especially where there is low turnover – that is, in strategic storage tanks and in commercial storage tanks in the quiescent bottom zone. These biofilms act as a source of recontamination for new fuel entering the tank and as inocula for downstream parts of the system. High fuel throughput and thorough cleaning of the walls of tanks when out of commission for any length of time will help to prevent the build-up of biofilms. In aircraft, problems are kept to a minimum by rigorous monitoring and use of permitted biocides when required.

Antimicrobials have been used to treat contaminated fuels and fuel systems since the early 1960s. Jet fuel systems were the first to receive such treatment, and frequent filter plugging and tank corrosion in marine systems led to increased use of biocides here also. Nowadays, microbiocides are employed in all grades of fuels. Used alone, however, they are rarely sufficient to correct severe microbial contamination problems. They are most effective when used as preventive treatments or together with cleaning of the system. Shock treatment of heavily contaminated surfaces will cause biomass to dislodge, and this may cause premature filter plugging. Because biocides are unlikely to kill microorganisms in heavy biofilms, the surviving cells can proliferate and accelerate the rate of system recontamination. Hence care must be taken in the selection of both the product and the treatment regime. The chemicals employed must have no adverse effects on fuel specifications or on the engines, pumps, and so forth, in the system. A range of biocides approved in the United States for use in fuels and fuel systems can be found in the California Environmental Protection Agency Office of Pesticides Programs on-line database: http://www.cdpr.ca.gov/docs/epa/epachem.htm

In the aircraft industry, the traditional biocides are ethylene glycol methyl ether (EGME) (an anti-icing agent with mild biocidal effects, used in military planes) and organoborinanes. A specially formulated isothiazolinone is now allowed by some companies and is more effective, if more expensive. A wider range of biocides is available for other fuels and includes compounds which are formulated to partition into either the fuel or the water phase. The best option is a biocide which dissolves in the fuel and partitions into the water phase at sufficient concentration to be effective. For treatment of storage tanks only, water-soluble biocides may be preferred, as the quantity of water to be treated is (generally) small, reducing costs, and the fuel phase is rapidly turned over. In laboratory simulation of biocide testing for the protection of diesel in storage tanks, an isothiazolone mixture and a quaternary ammonium compound were shown to be the most effective, whereas glutaraldehyde and a formaldehyde-releasing agent were active only at high concentrations, if at all. The same isothiazolone mixture was efficient against *H. resinae* in biofilms on steel surfaces, but much less so when SRB were also present.

The costs of biocide treatment obviously vary enormously, but they have been calculated at about US\$3 per ton of fuel for decontamination and US\$0.4–0.8 for continuous preventive treatment. An isothiazolinone biocide specially formulated for use in fuels was shown to maintain laboratory-scale diesel/water systems clean for 30 days when added once only (at the beginning of storage) at 0.1 ppm and for 400 days when used at 1 ppm. After 400 days, the systems treated with 0.1-ppm biocide showed increased microbial biomass, indicating the dangers of using sub-optimal concentrations of additives. For decontamination of a highly contaminated system, 10 ppm was necessary.

LUBRICANTS

Mineral oil lubricants are mainly composed of a variety of alkanes and naphthalenes together with small amounts (less than 25%) of aromatic compounds, whereas synthetic lubricants consist of methyl silicones and esters. Lubricants contain additives (Table 3.4) to improve performance, prevent corrosion, and reduce oxidation. There may be detergents, dispersants, acid neutralizers, and metal passifiers to prevent corrosion, making up perhaps 20% of the total oil formulation. Several of these additives have been found to be readily biodegradable. It is thus not surprising that lubricant formulations can provide a good source of microbial nutrients.

Problems associated with lubricants have been reported in two main areas, slow-speed marine diesels and the metal-working industry, both associated with the presence of water, either as a contaminant or an intimate admixture. In both situations the presence of large microbial infections has

---

Table 3.4. Typical oil additives

Metal soaps	Hindered phenols
Polyalkenyl succinimides	Aromatic amines
High molecular weight carboxylic acids	Alkyl phosphates
Metal dithiophosphates	Alkyl/aryl phosphates
Polyorganosiloxanes	

---

adversely affected the lubricating properties and increased the chance of corrosion and damage to machinery.

Marine diesel lubricants

Large marine diesel engines circulate vast amounts of oil, varying from 18 000 to 68 000 l (4000 to 15 000 gal), depending on whether the oil is used as a coolant. The oil acts as a heat sink for the engine, and it is necessary to pass it through a seawater cooler to maintain a temperature of 35 to 60 °C. Water from seepage and bilge is always present in the systems in small amounts (0.2% is considered an acceptable level), providing a source of inoculum. If a leakage occurs in a water jacket used to cool the pistons then large amounts of water, often containing nitrite corrosion inhibitors, can mix with the oil. Water levels of between 1% and 10% can be critical in encouraging microbial infection of the oil. The microflora present is predominantly Gram negative (including the nitrite reducers), but fungi may also be present, particularly *A. fumigatus*, which may be a primary colonizer. The symptoms which indicate an infection are as follows:

1. emulsification of the oil, which is probably the result of the production of microbial emulsifiers
2. uncharacteristic smell to the oil; reduction in pH from alkaline to acid
3. rust films and corrosion of the centrifuge bowl in the purifier system
4. marking of bearings by acid products; sludge accumulation and filter plugging.

Metal-working fluids

Millions of gallons of metal-working fluids (MWFs) are used each day in industry for cutting, milling, drilling, stamping, and grinding. They may be used as oils alone or mixed with water as an emulsion. Mineral-oil-based

MWFs were the most predominant lubricating/cooling fluids used during the first three quarters of the twentieth century. Soluble and semi-synthetic MWFs have now overtaken mineral oils as the most frequently used fluids. MWFs are grouped into four major classes: straight oils, soluble oils, semi-synthetic fluids, and synthetic fluids. The distinguishing feature is the amount of highly refined oil. *Straight, or neat, oils* are 60%–100% oil and contain free sulfur, phosphorus, chlorine, or fatty oils (extreme-pressure additives) to reduce metal-to-metal contact at high pressures. When they are used without dilution they do not present infection problems. *Synthetic fluids* contain no petroleum oils and may be water soluble or water dispersible. The simplest synthetics are made with organic and inorganic salts dissolved in water; these offer good rust protection and heat removal but usually have poor lubricating ability. Others may be formulated with synthesized hydrocarbons, organic esters, polyglycols, phosphate esters, and other synthetic lubricating fluids. They are often regarded as being *bioresistant*, or *biostable*, but microbial growth can occur, especially of yeasts and filamentous fungi. The two largest classes of lubricants, the *soluble oils and semi-synthetic fluids*, contain 5-85% oil. In general, soluble oil emulsions contain more than 50% mineral oil plus one or more of these components:

1. petroleum sulfonates
2. carboxylic acid soaps
3. non-ionic emulsifiers
4. corrosion inhibitors
    a. nitrites
    b. amines
    c. amides
5. extreme pressure additives
6. biocides
7. scents, dyes, antifoams
8. coupling agents, alcohols, and phenols.

The soluble oil and semi-synthetic cutting fluids represent the greatest potential for the growth of microorganisms because of the array of degradable additives and nutrients which they contain. In addition to the utilization of the oil component, it is the petroleum sulfonate emulsifying agent which is most readily utilized. This results in the breaking (cracking) of the emulsion into its component oil and water layers. A reduction in pH brought about by organic acid production by the microorganisms will

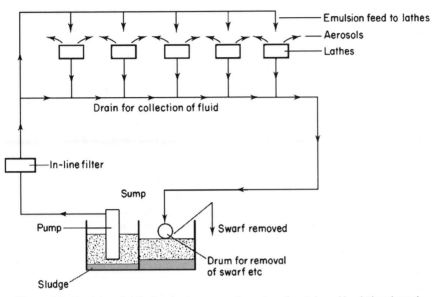

Figure 3.3. Emulsion distribution system around a series of metal-working lathes based on a centralized sump. Arrows along pipework indicate the direction of flow.

also contribute to emulsion cracking. To describe the range of causes and effects which are symptomatic of an infected oil emulsion, data from a fictional case history based on experience are presented below.

A factory manufacturing car components has an engineering shop which contains 100 lathes (Figure 3.3). The lathes are served by a centralized sump system containing 68 000 l (15 000 gal) of a mineral-oil-based soluble emulsion which does not contain any biocide. The emulsion circulates (3-h cycle) while the machines are in continuous use for 5 days/week. At weekends the pumps are turned off and the emulsion is allowed to drain back to the sump through open channels beneath the floor of the shop. The average temperature of the emulsion is a few degrees above ambient as a result of its use as a coolant at the cutting tool edge. The emulsion levels are periodically topped up with water from a well on site, which is stored in header tanks before use. There is some suggestion that, in the past, when the well was very low, it was the practice to draw water from an adjacent canal for use as a top up. It is also a very long walk across an open yard to the lavatories, especially at night in the rain! Symptoms began to develop which suggested that all was not well; the tool life was shortened, the fluid smelt unpleasant and turned a greyish colour, and the operators

Table 3.5. Bacteria and fungi isolated from metal-working emulsions

Bacteria

Achromobacter spp.	*Micrococus citreus*
Actinomycetes	*Nocardia* sp.
Aerobacter cloacae	*Proteus morganii*
Alcaligenes sp.	*Proteus vulgaris*
Bacillus cereus	*Pseudomonas aeruginosa*
Clostridium sp.	*Pseudomonas oleovorans*
Corynebacterium sp.	*Sarcina* sp.
Desulfovibrio desulfuricans	*Serratia liquefaciens*
Enterobacter cloacae	*Shewanella* sp.
Escherichia coli	*Staphylococcus* sp.
Escherichia freundii	*Streptococus pyogenes*
Escherichia intermedius	(haemolytic and non-haemolytic)
Klebsiella aerogenes	
Klebsiella pneumoniae	*Streptococus pneumoniae*
Listeria monocytogenes	

Fungi

Acremonium sp.	*Geotrichum candidum*
Aspergillus spp.	*Penicillium* spp.
Botrytis sp.	*Scopulariopsis* sp.
Candida lambica	*Torulopsis candida*
Cladosporium sp.	*Verticillium* sp.
Fusarium sp.	

complained of septic swarf cuts on their hands. On starting up the lathes on Monday morning, separated oil and water were delivered to the tool, followed closely by a strong smell of hydrogen sulfide (the 'Monday morning smell'), machined components began to corrode during storage, and filters became prematurely blocked.

The list of problems observed indicated a gross microbial infection of the emulsion which warranted attention. The causes are self-evident from the description of the system, and the result is the development of a large and diverse microbial flora. More than thirty species have been isolated from a single infected emulsion (see Table 3.5), ranging from free-living opportunist organisms to human pathogens. The ecology of infection is very complex and remains for the most part an undefined quantity. However, for practical purposes the course of infections have been documented so that early warning signals can be interpreted and remedial action taken. The course of an infection may be traced by use of the following

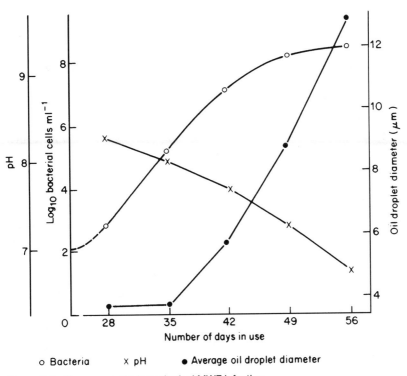

Figure 3.4. The course of a hypothetical MWF infection.

parameters: total aerobic bacterial count, presence of sulfate-reducing bacteria, pH changes, and increasing oil droplet size in the emulsion. The course of a hypothetical infection is shown in Figure 3.4.

In practice, the two indicators of a 'sick' emulsion are counts in excess of 10×10^6 bacteria/ml emulsion and a strong smell of hydrogen sulfide. Two other indicators are the formation of biofilms on pipework and walls of reservoirs and the separation of the oil and the water phases, which is often the result of a pH reduction caused by microbial activity (see Figure 3.4). Each industry has its own problems; those specific to the metal-rolling industry result from the high working temperatures of the emulsions (about 40 °C for steel and 50 °C for aluminium) and thus thermotolerant species have been isolated, including lipolytic strains in cases where vegetable oil formulations are employed as lubricants. Aggregates of bacteria may be burnt onto the steel during annealing, resulting in the possibility of rust spotting on storage. Aluminium rolling is particularly prone to sheet quality imperfections, and monitoring of the emulsion is particularly rigorous.

Problems arising from the use of vegetable oils in the synthetic yarn industry (see section on Impurities in Plastics) and of emulsions in the wire-drawing industry have been noted by the authors. In the drawing of copper wires, diamond dies drilled with holes of different diameters are used to draw out the wire to different gauge thicknesses. Undue wear on these dies quickly increases the diameters of the holes, and high microbial counts have been detected under such conditions. The link between the two is tentative at present as follow-up investigations were not possible in this case.

It is not within the scope of this book to discuss the health aspects of biodeteriogens in any detail. Pathogenic organisms have been isolated from infected emulsions, but their involvement with diseases in the workforce has seldom been proven. Although frankly pathogenic organisms such as *Salmonella, Staphylococcus,* and *Legionella* have been isolated from MWFs, most of the organisms are non-pathogens or opportunistic pathogens. Legionnaires' disease and Pontiac fever have both been noted to occur in MWF environments, where the infection is presumed to be caused by inhalation of MWF mists containing the responsible bacteria. Despite the frequency and severity of *Pseudomonas* infections in susceptible persons, healthy adults with intact immunity are rarely affected. No reports have been published of work-related *Pseudomonas* infections in metal workers. One study of a worksite with a demonstrated viable count of 1×10^8 colony-forming units/ml of MWF showed no evidence of *Pseudomonas* colonization of the workers' respiratory tracts, even though the organisms were cultured from the emulsion. Some interest has focused on the possible involvement of microbial antigens in recent clusters of hypersensitivity pneumonitis among workers exposed to MWF aerosols, and aerosolized endotoxins (bacterial lipopolysaccharides) are suspected causative agents of occupationally related respiratory problems (e.g., chronic bronchitis, decline in pulmonary function, asthma) among workers exposed to MWF aerosols. However, the evidence so far is not strong.

A wide variety of biocidal substances have been used in attempts to prevent, retard, or remove microbial contamination in MWFs. Some of these are shown in Table 3.6. These substances themselves may be hazardous to human health, for example, formaldehyde releasers (usually more effective against bacteria than fungi), such as tris (hydroxymethyl) nitromethane

---

Table 3.6. Some biocides used in MWFs

Tris(hydroxymethyl)nitromethane
Hexahydro-1,3,5-tris(2-hydroxyethyl)-S-triazine
Hexahydro-1,3,5-triethyl-S-triazine
1-(3-chloroallyl)-3,5,7-triaza-1-azonia adamantane chloride
4-(2-nitrobutyl)morpholine-4,4'-(2-ethyl-2-nitrotrimethylene) dimorpholine
O-phenyl phenol
Sodium 2-pyridinethiol-1-oxide
1,2-benzisothiazolin-3-one
5-chloro-2-methyl-4-isothiazolin-3-one+2-methyl-4-isothiazolin-3-one
6-acetoxy-2,4-dimethyl-*m*-dioxane
2,2-dibromo-3-nitrilopropionamide
p-chloro-*m*-xylenol
2-*n*-octyl isothiazolin-3-one
3-iodopropynyl butyl carbamate

---

and hexahydro-1,3,5-tris(2-hydroxyethyl)-S-triazine. Formaldehyde is an airways irritant and a recognized cause of occupational asthma. Studies have also suggested that exposure to certain antimicrobial agents can cause allergic or irritant contact dermatitis. Legal limits are now in place to minimize these risks (see Chapter 6). Other problems are the development of biocide-resistant strains and the formation of biofilms in the system, resulting in a source of recontamination. In a model experimental MWF circulating system, after 6-weeks operation, the aerobic and anaerobic bacteria and fungi in the soluble oil fluid containing a triazine biocide reached approximately 10^{10}, 10^{10} and 10^5/ml, respectively, and the equivalent biofilm populations were 10^{10}, 10^8, and 10^5/cm^2. Physical disruption of the biofilm in another system produced an increase of up to $20\times$ in the planktonic population.

Biodeterioration of plastics and rubbers

Plastics and synthetic rubbers are essentially a modern group of materials, having been commercially in use in any quantity for only about 60 years. The generic term plastic covers a wide range of diverse structures and indeed does not limit itself to the polymer, but includes formulations, composites, and copolymers. Thus, when susceptibility to attack is considered, it is important to know exactly what the plastic represents in terms of its overall composition. Plastics may contain a wide variety of additives, and they may have residual processing aids adhering to their surfaces; all of these substances affect susceptibility. Rubber is a natural product, and we would thus expect some microorganisms to be able to utilize it in its raw form (latex), which contains proteins and carbohydrates held in an aqueous phase, and this is indeed so. Rubber is processed to remove these impurities and improve its performance properties. Formulation additives in rubber may also encourage microbial growth.

Physical factors are also important in determining the susceptibility of plastics and rubber. Hydrophobicity may restrict the intimate contact required between the extracellular enzymes of the biodeteriogen and the substrate. The surface texture of the product may encourage attachment if it is roughened, and increase the rate at which colonization can occur over a smooth impermeable surface. The hardness of the material is important where insect and rodent attack is concerned. Shape and form may also be of significance in rodent attack, as rodents need to gain purchase on an object to gnaw it. Large-diameter pipes and cables are more resistant than "bite-size" small ones. The way in which the plastic is presented to the environment is important. For example, stress on a susceptible plastic enhances attack by microorganisms and shortens its service life. Exposure to aqueous and soil environments will have similar effects. Such factors are poorly understood and seldom evaluated (see Chapter 5); thus they are often misinterpreted.

Published literature concerning biological deterioration of polymers often lacks specific information on composition details, making interpretation of the data difficult, and data collected from in-service failures are often either confidential or misinterpreted by on-site inspectors and subsequently destroyed. Samples taken from a site may be badly stored so that subsequent examination of the causative organism is made more difficult. It is thus difficult in some cases to make definite conclusions on the performance of the product in service.

The main polymer groups are described in the following sections and their main repeating backbone structural units illustrated. The majority of these structures represent extremely non-reactive chemical groups, making up high-molecular-weight, saturated, unbranched polymers, which, on curing, cross-link their structures. However, plastics based on modified or regenerated cellulose and the introduction of carbonyl, ester, amide, alcohol, and urethane groups all present potential sites for enzyme-mediated reactions. The only groups for which published evidence implicates direct attack on the structure are natural rubber, the regenerated celluloses, the regenerated proteins, the polyesters, the polyurethanes, and the nylons. It is interesting to note that the structures of these polymers are either derived from or incorporate naturally occurring moieties. This suggests that the microorganisms do not need to synthesize novel enzyme systems to degrade these substances. Of course, the secondary and tertiary structures acquired during the processing and curing of these polymers may adversely affect the ability of large enzyme molecules to reach the susceptible group sites and catalyse degradation.

NATURAL AND SYNTHETIC RUBBERS

Rubber in its natural state consists of a long chain of repeating *cis*-1,4-isoprene units (see Figure 3.5) with an average molecular weight of 2×10^6 and a protein content of between 2% and 3.5%. It is subject to oxidation, and it is thought to be this non-biological precursor which encourages microbial utilization of the oxidation products. Pale crepe, used for example in the manufacture of surgical gloves, is the purest form of latex. It contains up to 15% non-rubber constituents such as proteins, lipids, and carbohydrates. It can absorb up to 15% water in its unvulcanized state, making it particularly susceptible to biodeteriotation when stored in damp conditions. Latex undoubtedly biodegrades naturally as evidenced by the low residual levels in the soil of rubber plantations, where considerable amounts are deposited with falling leaves each year.

Vulcanization improves the mechanical properties, working temperature range, and chemical resistance of natural rubber by direct cross-linking of the chains through carbon atoms by use of radiation/peroxide agents or by means of a sulfur group if chemical accelerators are used. Some accelerators such as mercaptobenzothiazole, tetramethyl thiuram disulfide, and sodium dimethyl dithiocarbamate exhibit biocidal activity and are used as preservatives in solvent-based products such as plastics, adhesives, and paints for which dry-film fungal protection is required. The reduction in water absorption and the cross-linking also serve to reduce

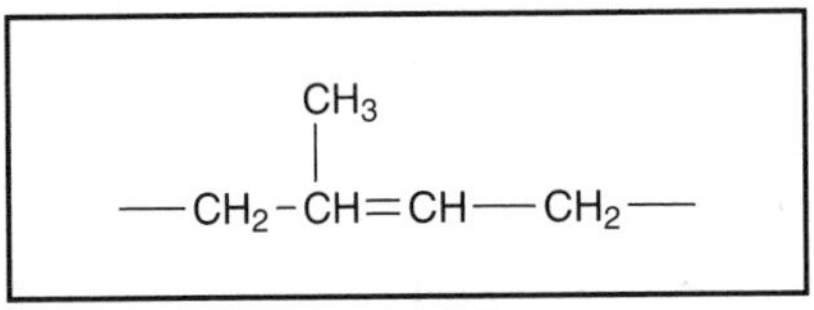

Figure 3.5. 1,4-isoprene unit.

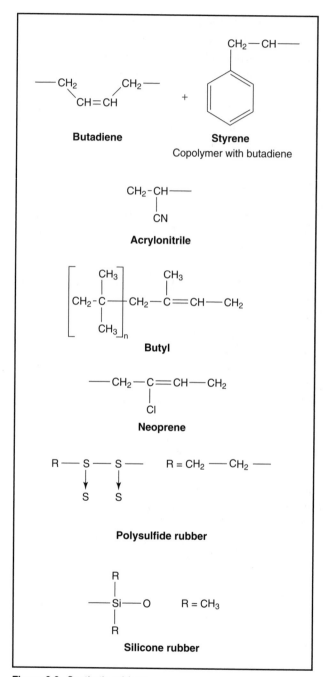

Figure 3.6. Synthetic rubbers.

microbial susceptibility. However, weight-loss experiments in both mixed and pure culture systems have shown that, in time, significant degradation will occur. Studies with rubber particles and vehicle tyres in soil and with a pure culture of *Nocardia* sp., respectively, have suggested that a biodegradation process might be feasible for the removal of waste rubber. It is thought that the degradative pathway involves bacterial metabolism of the polyisoprenoid compounds geraniol, citronellol, and farnesol. The inducible enzyme, geranyl coenzyme A carboxylase, has been identified in the degradation of isoprenes by *Pseudomonas aeruginosa*.

A range of synthetic rubbers has been developed to improve the properties of natural rubber (Figure 3.6). Some of these rubbers, such as isoprene, butadiene, and chloroprene, are similar in structure to natural rubber, whereas others, such as butyl, acrylonitrile, and styrene, have additional functional groups incorporated into the isoprene moiety. In contrast, silicone and polysulfide rubbers have backbones based on silicone.

There are very few reports on the degradation of these rubbers, with the exception of the styrene–butadiene copolymer. A *Nocardia* sp. has been shown to be capable of the degradation of butadiene through the stereospecific epoxidization of the $C = C$ bond and subsequent oxidation through pyruvate to acetate.

REGENERATED AND MODIFIED CELLULOSES

Regenerated celluloses, such as cellulose acetate and cellulose nitrate, were some of the earliest plastics to be used commercially. The substitution of hydroxyl groups for either acetate or nitrate (Figure 3.7) confers increased resistance over non-substituted cellulose. This has also been found for cellulose ethers (Figure 3.7) such as carboxymethyl and hydroxyethyl celluloses which, depending on the degree of substitution and the substituting group, are invariably less susceptible to cleavage at the β-1-4 linkage than is their natural parent structure. This is not to say that these forms do not present problems. Indeed, the loss of viscosity in emulsion paints during storage is often due to enzymatic attack on the cellulose ether thickener; large changes in viscosity can be effected by small numbers of random points of attack. The fabric known as rayon is a cellulose derivative which can easily be shown to be utilized by microorganisms. Strips of rayon buried in a microbiologically active soil for 1 month completely lose tensile strength. Cellophane, once widely used as a wrapping, will support excellent microbial growth, particularly that of fungi. Many

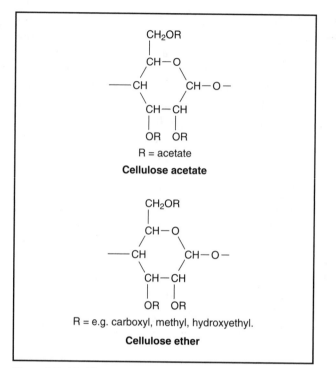

Figure 3.7. Modified celluloses.

microbial ecologists who studied the cellulolytic activities of the soil in the past used cellophane to select out cellulolytic species and to study their growth characteristics in pure culture in the laboratory. Because of this susceptibility, the use of cellulose derivatives in packaging has given way to more inert polymers which often exhibit superior mechanical, physical, and chemical properties. Cellulose, as we have seen in Chapter 2, is still in great demand as a raw material. There has, however, been a resurgence in the use of novel derivatives from cellulosic wastes. This has been due to both a need to make more efficient use of the vast amounts of fibrous waste which accumulate from agricultural and food processing and the advantages of using a renewable resource versus one which is essentially one-way. Thus furfural derivatives have been produced from hemicelluloses in waste cereals, and these have been introduced into insulating foam in conjunction with phenolic resins. These appear to be resistant to microbial attack under laboratory conditions.

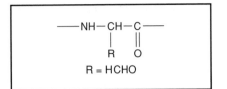

Figure 3.8. Unit structure of casein–formaldehyde.

REGENERATED PROTEINS

Some of the earliest commercially produced plastics were regenerated proteins, produced from casein which had been extracted from skimmed milk and hardened with formaldehyde. This casein–formaldehyde was used for electrical fittings such as light switches. The unit structure is shown in Figure 3.8, where the R group is formaldehyde. These plastics are subject to water absorption and hydrolytic attack at the carbonyl sites, and their susceptibility to microbial attack supports this observation. Urea formaldehyde is a similar compound, which is used as a bonding agent for wood particleboard and compressed paper in domestic worktop laminates. The properties of the polymers in this class, such as their brittleness and hydrolysis in contact with water, have limited their use in industrial products.

POLYETHYLENES

Polyethylene comprises a highly polymerized saturated $C — C$ backbone with occasional short branches (2–3 carbons in length) which are terminated by a methyl group (see Figure 3.9), rendering it extremely recalcitrant to microbiological attack. High-density polyethylenes (HDPEs) have less than five branches per 1000 carbon atoms and low-density polyethylenes (LDPEs) have between twenty and thirty. It is generally recognized that oxidation (whether photo- or chemical) is a necessary precursor to any biological degradative process. It has been found that photo-oxidation of ^{14}C-labelled polyethylene increases biodegradation in the soil. Sam-

$$[-CH_2-CH_2-]_n$$

Figure 3.9. Repeating structure of polyethylene.

ples of polyethylene subjected to 42-day photo-oxidation and then buried for 10 years showed a tenfold (2%) increase in degradation over non-photo-oxidized samples. It is postulated that the subsequent biodegradative pathway proceeds by β-oxidation. Thus we might predict a very slow mineralization process even in the presence of abiotic oxidation. The hydrophobicity and incorporation of antioxidants into polyethylene further reduce the degradation rate.

$$HO(CH_2CH_2O)_nCH_2CH_2OH$$

Figure 3.10. Structure of PEG.

Polyethylene glycols (PEGs), in contrast, are well characterized in respect to their microbial degradation. PEGs in molecular weights from 400 to 20 000 (Figure 3.10) are metabolized by a range of gram-negative bacteria through successive oxidation of the terminal alcohol group to an aldehyde and monocarboxylic acid, followed by cleavage of the ether bond. Three PEG–dehydrogenases have been identified as responsible, resulting in the reduction of one glycol unit on the PEG chain.

Polystyrene, polyvinyl chloride (PVC), polyacrylonitrile, polyacrylate, and polyvinyl acetate all contain pendant groups from the repeating saturated CH_2 backbone. The literature records biodegradability and mineralization of styrene, acrylonitrile effluents, and soluble polyacrylate and polyvinyl acetate. However, as rigid plastics these polymers are extremely recalcitrant.

POLYESTERS

Polyesters are esters of an organic acid and a polyol. The acid is generally dibasic and the polyol is a dihydroxyl alcohol. The susceptibility of the polyester to biodegradation depends primarily on the type of acid used: adipates, sebacates, and caproates are generally recognized as producing degradable polyesters, whereas those derived from phthalates, toluene sulfonic acid, aromatic hydrocarbons, and polycarbonates are more resistant to degradation by microorganisms. Initial degradation of susceptible polyesters is due to the unstable ester bond, which is hydrolyzed in the presence of hydrolase enzymes. Unbranched polyesters are more susceptible than branched types, even when the molecular weight of the latter is much lower. As an example in a related area, the insertion of methyl groups

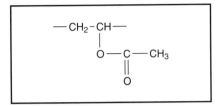

Figure 3.11. Polyvinyl acetate repeating unit.

into the hydrocarbon dodecane results in no growth of a mixed culture of standard test fungi, whereas dodecane alone supports excellent growth under the same conditions. Ester linkages in a pendant position on the main chain, for example, polyvinyl acetate, are not attacked (see Figure 3.11). In general, the higher the melting point the lower the biodegradability tends to be. It has been reported that the enzymatic hydrolysis of the aliphatic polyester poly(tetramethylene succinate), with a high melting point (113 °C), is lower than for poly(ε-caprolactone), with a melting point of 62 °C. The chain length of the unbranched polyol also appears to have an effect on the growth of fungi. A linear alkane alcohol with a chain length of between C3 and C12 incorporated into an ester with sebacic acid stimulated growth of *A. versicolor*, whereas above C12 growth was reduced. The alcohol alone did not support growth of the fungus. Butan-1,4-diol is commonly used as the polyol constituent in many applications, and several studies have shown that, in addition to the chain length, the position of the hydroxyl groups has an effect on the growth of fungi. The 1,4 positions appear to result in a more susceptible ester than the 1,2 or 1,3 positions. This evidence points to neighbouring groups influencing the enzymes responsible for the reactions. Susceptible polyesters such as polycaprolactone and polybutylene adipate are incorporated chemically into the structure of some polyurethanes and here the effects of the adjacent polyurethane groups have to be considered. These compounds are discussed more extensively in the section on polyurethanes.

Although polycarbonates have greater resistance to hydrolysis than do polyesters, microorganisms capable of the formers' degradation have been isolated. More interestingly, a strain of *Amycolatopsis* sp., a polyester-degrading bacterium, has been found to be capable of complete mineralization of a copolymer of tetramethylene succinate and tetramethylene carbonate (Figure 3.12). In contrast, the polytetramethylene carbonate showed little degradation.

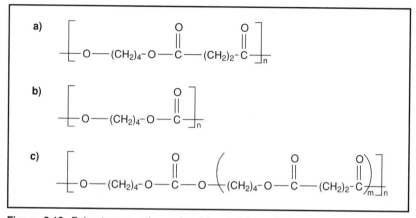

Figure 3.12. Polyester repeating units: (a) poly(tetramethylene succinate), (b) poly-(tetramethylene carbonate), (c) copolymer of (a) and (b).

Of the resistant polyesters, polyethylene terephthalate or PET (trade names, Terylene and Dacron) is the most widely used (Figure 3.13). It is resistant to hydrolysis and to degradation by microorganisms. Small amounts of growth on Terylene have been reported, but this may have been due to pigments, low-molecular-weight impurities, or processing aids. The degradation of dimethylterephthalate by *A. niger* has been observed in vitro and a tentative degradation pathway proposed. Apart from this, there is no evidence to suggest that PET is a susceptible polymer.

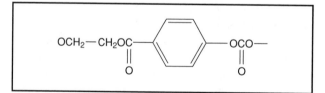

Figure 3.13. PET structure.

POLYURETHANES

The term polyurethane describes a wide range of polymeric materials, including flexible and rigid foams and elastomers. They are used in the manufacture of paints, adhesives, furniture, synthetic leathers, electrical cables, automotive components, and medical devices. In their early development, the predominant chemical group present in the chain was

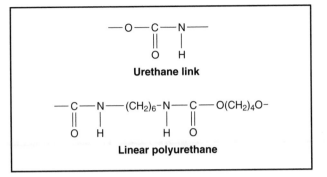

Figure 3.14. Basic polyurethane structure.

the urethane group, which gave rise to the generic term polyurethane
(Figure 3.14). The rapid expansion of this group of materials over the past
30 years has resulted in the introduction of structurally important groups
such as esters, ethers, and ureas into the structures of polyurethanes. The
polyurethanes are a prime example of a polymer which contains susceptible functional groups, known to fail in service, which has been extensively
studied in recent years. Because of their unique properties (good abrasion,
water and oil resistance, good adhesion, elasticity, and high strength), the
variety of uses of polyurethanes is becoming increasingly diverse such that
they may involve close contact with microbial communities in the soil and
aqueous environments such as the sea, the human body, and sewerage systems. The simplest polyurethane is formed by the combination of a polyol
with a highly reactive di-isocyanate, resulting in a rigid polymer of regularly
repeating urethane linkages (Figure 3.14). Such a polymer has limited use
as a rigid insulating foam. The incorporation of a polyether or polyester
produces an elastomeric prepolymer which can be extended with either a
diol or a diamine to produce a long chain product. Cross-linking can occur
(Figure 3.15), and the use of triols can produce branched structures which
result in a thermoset plastic in which the cross-linking remains stable and
the polymer cannot be remoulded by heating. The product may be cured
by high temperature, resulting in a totally reacted product with segmented
structure consisting of a soft (elastic) segment made up of the polyester
or polyether chain and a hard segment which is the isocyanate moiety
(Figure 3.16).

Polyester-containing polyurethanes have been found to be more susceptible to attack and experience more in-service failure than those based
on polyether (Figure 3.17). This is due to the susceptibility of the ester to

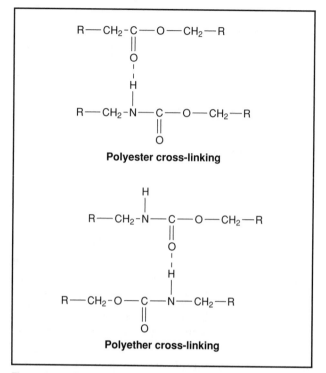

Figure 3.15. Cross-linking in polyurethanes.

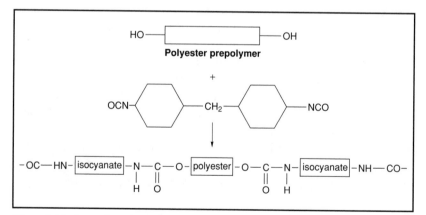

Figure 3.16. Formation of a segmented polyester polyurethane.

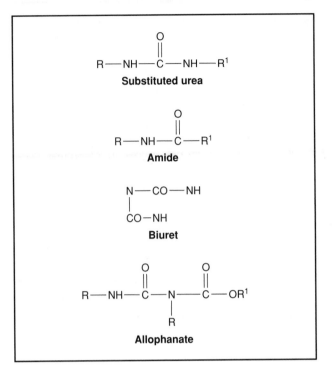

Figure 3.17. Polyurethane groups.

hydrolysis, which a large number of microbial enzymes (the hydrolases) catalyse: the effect can be demonstrated with thin films of a susceptible polyurethane within 14 days of burial in an active garden soil. The ether bond is susceptible to oxidation (it is less stable in hot dry atmospheres than the ester), but appears to be resistant to microbial attack. This conclusion is based on observations of growth on either commercial products or laboratory-synthesized polymers. More recent work postulates that microorganisms may also be involved in the opening of the aromatic rings of the di-isocyanate and direct hydrolysis of the urethane, amide, and urea groups. However, this seems to be related to the amount of hydrogen bonding (cross-linking) within the hard segment and the shielding of these groups through secondary and tertiary (crystalline) structures. This would mean that ether-based polyurethanes might be at risk under certain conditions. Observational data from growth of mixed fungal cultures on a range of ether- and ester-based polyurethanes have shown that 'moderate' amounts of growth can be obtained from polyethers when incorporated into aromatic di-isocyanates. The aliphatic di-isocyanate, 1,6-hexame-thylene di-isocyanate, gave consistently lower growth ratings for both polyether- and polyester-based systems. The literature on protease activity (another hydrolase) suggests that these enzymes preferentially attack a site close to an aromatic ring in other structures; this may explain the more resistant 1,6-hexamethylene di-isocyanate-based systems.

The bulk of the work in recent years has concentrated on isolating and identifying the enzymes responsible for the degradation of polyurethanes. Both proteases (papain and urease) and lipases/esterases isolated from aerobic and anaerobic bacteria have been shown to actively degrade the urethane linkages and polyester segments, respectively, resulting in loss of structural integrity. Further work has revealed the presence of both extracellular and membrane-bound heat-stable (10 min at 100 °C) enzymes in *Pseudomonas fluorescens, P. chlororaphis,* and *Comamonas acidovorans* with molecular weights in the range 30 000–65 000 which are capable of degradation of the polyester segment. Such is the interest in these enzymes for their potential use in polyurethane waste removal that time has been spent cloning some of the PUase genes and expressing them in *Escherichia coli.*

POLYAMIDES

These polymers are generically referred to as nylon and are synthesized by the self-condensation of ε-amino acids or their lactams or by condensation of diamines with dicarboxylic acids. Nylon 6 (prepared from caprolactam) and Nylon 11 (prepared from ε-amino undecanoic acid) are examples of the first reaction, whereas Nylon 6.6 (hexamethylene diamine and adipic acid) and Nylon 6.10 (hexamethylene diamine and sebacic acid) represent the second reaction (Figure 3.18). This yields repeating units with molecular weights between 100 and 200 with polymers for commercial use of between 10 000 and 100 000. The literature reports that nylons are for the most part very resistant to biodegradation. They are more likely to be colonized by staining fungi which grow on surface contamination. An example of this is the pink staining of nylon parachutes by *Penicillium janthinellum.*

Information concerning microbial degradation of polyamides is restricted to copolymers of either glycine or serine and ε-amino caproic acid (Figure 3.19). These copolymers were developed for their enhanced

—NH(CH$_2$)$_5$CO—	Nylon6
—NH(CH$_2$)$_{10}$CO—	Nylon 11
—NH(CH$_2$)$_6$NH—CO(CH$_2$)$_4$CO—	Nylon 6.6
—NH(CH$_2$)$_6$NH—CO(CH$_2$)$_8$CO—	Nylon 6.10

Figure 3.18. Polyamides.

biodegradability. Their hydrophilic nature and water solubility, in contrast to polyglycine (Nylon 2) and polycaprolactam (Nylon 6), both of which show good resistance to microbial attack, were thought to be responsible for the ability to support microbial growth. Oligomer units of nylon (ε-amino hexanoic cyclic dimer) have been found to be biodegradable, and the gene responsible for expressing the enzyme has been identified. Our experience is that nylon is not prone to structure-related failures.

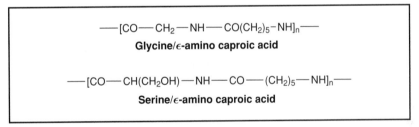

Figure 3.19. Two types of Nylon 2 and 6.

ADDITIVES

Plastics and rubbers may contain a variety of additives (Table 3.7) which confer various properties on the final product. Very little work has been carried out on these additives to determine their nutrient or biocidal properties, with the notable exception of plasticizers, which have been examined more closely because of their important plasticizing function in the microbiologically inert polymer PVC.

Table 3.7. Additives used in plastics

Pigments	Plasticizers
Antimicrobial agents	uv stabilizers
Metal catalysts	uv absorbers
Polymer flow controllers	Hydrolysis stabilizers
Releasing agents	Antioxidants
Lubricants	Flame retardants
Accelerators	Inert fillers and extenders

A list of the chemical species of plasticizers is given in Table 3.8 together with a qualitative indication of their susceptibility to biodeterioration. Plasticizers based on adipates and sebacates are generally accepted as being susceptible to biodeterioration, resulting in loss of plasticity of the PVC and eventual localized cracking around the microbial (usually fungal) colony. Other types of plasticizer are resistant when presented as the sole carbon source, but the presence of other organic

Table 3.8. Examples of commercial plasticizers and the ability of fungi to utilize them

Adipates	Colony diameter[a] (cm)	Sebacates	Colony diameter (cm)
Dihexyladipate	2.8	Dimethyl sebacate	1.0
Dibutoxyethyl adipate	1.5	Dibutyl sebacate	4.4
Dicapryl adipate	4.6	Doctyl sebacate	2.6
Dioctyl adipate	1.0	Di-iso-octyl sebacate	1.6
Di-iso-octyl adipate	1.1	Dibenzyl sebacate	5.5
Dinonyl adipate	0.4	Polypropylene sebacate	5.9
Phthalates		**Stearates**	
Dibutyl phthalate	0.4	Butoxyethyl stearate	6.2
Dioctyl phthalate	0.1	Butylstearate	2.2
Di-iso-octyl phthalate	0.0	**Ricinoleates**	
Dinonyl phthalate	0.0	Methyl ricinoleate	6.0
Didecyl phthalate	0.0	Zinc ricinoleate	6.8
		Butyl ricinoleate	6.1

[a]Numbers are averaged colony diameters (cm) for twenty-four fungi individually grown for 14 days at 29 °C on a mineral salt agar containing the plasticizer.

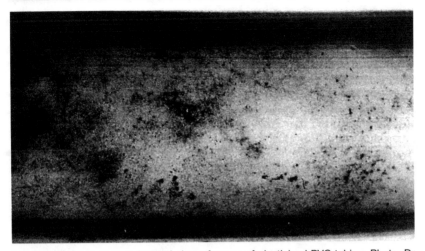

Figure 3.20. Pink stain on internal circumference of plasticised PVC tubing. Photo: Dr K. J. Seal.

nutrients in the medium stimulates their utilization. Whether this is a cometabolic or an induced enzyme process is not known. Plasticizers are esters which are hydrolysed by lipases/esterases found throughout the fungal and bacterial orders; thus plasticized PVC can support a wide range of fungi, often producing a stain as the first recognizable effect (Figure 3.20). The most well-known stain of plasticized PVC is the 'pink stain' caused by the presence of *Streptomyces rubrireticuli*. This organism forms part of a screening procedure for evaluating potential biocides (see Chapter 5).

The use of additives such as antioxidants and UV stabilizers not only serves to reduce the effects of the physical environment on the service life of the polymer, but also retards the changes often necessary to allow microbial growth to be initiated. Conversely, antihydrolysis agents used, for example, in polyester polyurethanes, are not effective in preventing enzyme-catalysed hydrolysis. They are designed to mop up free acid which can become autocatalytic, but cannot prevent the enzyme-mediated reaction from proceeding. Comparisons between identical polymers with and without antihydrolysis agents have shown that attack can occur to the same extent. Very little work has been done to examine the susceptibility of many of these additives. What little information is available indicates either a passive or, in the case of certain metallic catalysts, an active resistance to microbial attack.

IMPURITIES

The impurities found in plastics range from uncured or unreacted low-molecular-weight precursors and catalysts to residual mould-release agents and lubricants used during the manufacturing process. Work carried out on low-molecular-weight polyethylene residues present in a high-molecular-weight polyethylene showed that growth of soil microorganisms was arrested when the former low-molecular-weight residues were removed by a suitable solvent. The authors have observed the effects of residual vegetable oil, used as a lubricant in the production and weaving of polypropylene. Storage of rolls of the fabric in a warm damp environment resulted in the whole of the surface being covered in a mass of fungal colonies which, although not affecting the tensile strength, caused extensive staining.

Glass

The biodeterioration of glass is poorly understood, although it is an economic problem in the humid climates of the world. Glass provides a surface on which free silicon bonds can react with hydroxyl, methyl, and amino groups. Fungi, bacteria, and algae produce organic acids which can, through ion exchange, remove sodium ions from the surface, resulting in etching. For example, glass lenses in optical instruments may be severely damaged as a result of bad storage in the tropics. Fungi are the main agents, growing across the surface of the lens (Figure 3.21), perhaps utilizing a susceptible coating or contaminants which may be present on or around the lens. Species of xerophilic fungi are mainly responsible for the initial attack. Etching follows, possibly as a result of organic acid production and chelation. Some fungi form water-soluble complexes with metals present in the silicate, accelerating the rate of attack. Lichens have been observed colonizing the exterior surfaces of stained glass windows in Swedish churches. The lichens cause a slight change in the colour and extensive pitting of the glass surface. The metal oxides used as the stains did not appear to prevent growth, except on the yellow and grey panes where silver salts and iron oxides containing borax were employed. Cyanobacteria have also been shown, in laboratory experiments, to cause biofilm formation and pitting on glass. The pattern of biopitting produced by cyanobacterial and fungal cultures was very similar to the pits observed on antique and medieval glasses.

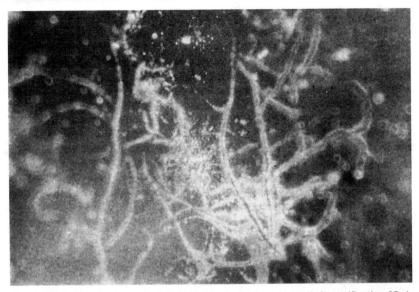

Figure 3.21. Etching of spectacle lens that is due to fungal growth (magnification 65×). Photo: Dr K. J. Seal.

Paints

Paints are used as protective coatings to prevent the environmental weathering of materials and to provide a decorative finish. In their simplest form they comprise a binder, a pigment, and a solvent. Binders produce the film as the paint dries by cross-linking (curing) and are normally based on either vinyl acetate, vinyl chloride, acrylate, or styrene polymer lattices (emulsions). The pigment, which provides the colour, can be either mineral or organic based. The solvent is either water or hydrocarbon based and assists in the application and drying properties. Modern paints also contain surfactants, rheology modifiers (e.g., cellulose ethers), and extenders (e.g., clays or calcium carbonate). Many of the raw ingredients provide nutrients for bacteria and fungi, and thus paints are susceptible to biodeterioration. They are vulnerable while still in the can and as an applied paint film. Although both solvent-based and water-based systems are susceptible, in-can problems are almost solely confined to water-borne emulsion paints, whereas spoilage of the film occurs in both types. The move towards paints with lower levels of volatile organic compounds (VOCs) to satisfy atmospheric emission controls has resulted in an increase in the level of microbiological susceptibility, even in emulsion systems in which glycols

Table 3.9. Bacteria isolated from liquid paint

P. aeruginosa
P. putida
P. fluorescens
Alcaligenes spp.
Proteus sp.
Citrobacter freundii
Corynebacterium spp.
Bacillus spp.

normally used have been restricted. Reductions in residual monomers from the binder which inhibit microbial growth have also contributed to increased susceptibility.

In the liquid state, the paint may be colonised by a range of gram-negative and gram-positive bacteria (Table 3.9), particularly the spore formers (*Bacillus* spp.) in the latter case. The pH of most paints is in the range 8–9.5, and this favours the pseudomonads, which are the most commonly encountered group, comprising at least 75% of isolates from spoilt paint. The main effects of bacterial attack are gas evolution, the production of 'off' odours, loss in viscosity of the paint (and to a lesser extent, increase in viscosity), emulsion splitting (Figure 3.22), pH change, discolouration, and the growth of bacterial or fungal colonies on the surface of the paint. Any one of these symptoms will render the paint unsaleable. The problem which has received most attention over the years has been loss in viscosity.

Maintaining the viscosity of an emulsion paint is very important as this affects its application properties; the most common thickeners used are cellulose ethers, such as hydroxyethyl cellulose, and these are subject to enzymic hydrolysis in the presence of both bacterial and fungal cellulases. The degree of resistance to hydrolysis is related to the degree of substitution of the hydroxyethyl groups. A regular pattern of substitution on each glucose unit of the polymer confers a high degree of resistance to biodeterioration. The group involved in the substitution is also of importance: Hydroxypropylmethyl cellulose is more resistant than hydroxyethyl and carboxymethyl cellulose to enzyme-induced viscosity losses. However, small amounts of cellulase (0.1 ppm) can cause significant (20%) decreases in viscosity. Cellulases are extracellular enzymes (see Chapter 2) and, as such, are capable of hydrolysis of the substrate even when there

Figure 3.22. 'Split' emulsion paint. Photo: Dr K. J. Seal.

are no microorganisms present. Cellulases are relatively stable enzymes which will remain active in a can of paint over a period of storage; they are inactivated by heat and poisoned by mercury-containing compounds. Such organomercurial biocides are no longer employed as in-can preservatives as a result of environmental and toxicological legislation. Viscosity losses can also occur as a result of residual oxidizing or reducing agents in the paint formulation as a result of carryover in the binder. Single agents, for example, hydrogen peroxide and sodium formaldehyde sulfoxylate, or redox systems, comprising an oxidizing and a reducing agent, are used as initiators for polymerization of the latex monomer. To reduce residual monomers in the paint (reduce VOCs) redox chemicals are also added at the end of the polymerization process. However, the extent to which viscosity loss occurs over time is not as great as for enzyme-mediated degradation, and this may be used as an early indicator of the cause. It should also be noted that residual redox agents can destabilize some preservative molecules. The isothiazolinones are particularly prone to inactivation by redox agents (see Chapter 6).

Fermentation of the cellulose ether by bacteria results in gassing. Gassing causes cans to bulge, eventually blowing off their lids. The source of the contamination may reside in the raw materials, the water supply, or the equipment used. Good housekeeping, hygiene, and monitoring systems can go much of the way to keep this kind of infection under control.

The polymer latex (emulsion), which may comprise as much as 40% of the formulation of a premier quality emulsion paint, is potentially a major source of both nutrients and contamination. Table 3.10 shows the range of bacteria and fungi which can be isolated from polymer emulsions. A latex will typically contain, in addition to the polymer, surfactants/colloids,

---

Table 3.10. Microorganisms isolated from polymer latices

P. aeruginosa
P. fluorescens
P. putida
P. (Burkholderia) cepacia
P. stutzeri
Alkaligenes faecalis
Aeromonas hydrophila
Achromobacter sp.
Citrobacter freundii
Acinetobacter sp.
Escherichia coli
Proteus vulgaris
Micrococcus luteus
Bacillus subtilis
Geotrichum candidum
Fusarium solani
Cladosporium herbarum
A. niger
Alternaria alternata
Candida albicans
Rhodotorula rubra
Torulopsis spp.

antifoams, residual monomer, initiators, buffering agents, and metal catalysts. The pH will range from 4.5 (vinyl acetate) to 9 (styrene acrylic), encouraging the growth of many types of fungi and bacteria. Typically, at acid pH, the major contaminants are either *Geotrichum candidum* or *Acinetobacter* sp., whereas at alkaline pH those pseudomonads found in paint predominate.

Microbial growth on paint films occurs on both the external and the internal surfaces of buildings and can occur on both emulsion and alkyd gloss paints. Externally, the problems are greater in the tropics, where the fungi and cyanobacteria cause defacement of painted structures. On internal surfaces the environmental conditions may promote a larger range of microorganisms, and the predominant flora comprises fungi and actinomycetes. Determining whether a paint film is being actively attacked is complicated by the fact that one or more of three processes may be in operation:

1. The organism may be directly utilizing a component of the paint;
2. It may be living on surface dirt and releasing a pigment which is absorbed by the paint;
3. It may be colonizing the substratum, e.g. wood, and erupting through the film.

A range of fungi, cyanobacteria, and algae have been observed on paint films (Table 3.11). Perhaps the most important are those fungi which produce dark-pigmented spores such as species of *Aureobasidium*, *Cladosporium*, and *Alternaria* and the dark-coloured cyanobacteria (often the coccoid types).

The algae and cyanobacteria are particularly important as aesthetic deteriogens of paint films. Although they are extremely prevalent in tropical and sub-tropical jungle climates, their presence is also important in the temperate Northern Hemisphere where damp yet cold conditions apply. Surveys carried out in both South-East Asia and South America demonstrate the ubiquitous nature of the range of algal/cyanobacterial species found on painted surfaces.

The success of these genera in colonizing paint films is due to a combination of tolerance to high temperature, periodic desiccation, and UV radiation. Other microorganisms form a succession of colonization, and it thus seems likely that an ecosystem exists on the film surface which leads to the observed deterioration effects.

Table 3.11. Fungi and algae isolated from paint films

Fungi	
A. alternata	Paecilomyces variotii
A. flavus	Penicillium expansum
A. versicolor	Penicillium purpurogenum
Aureobasidium pullulans	Pestalotia macrotricha
C. herbarum	Phoma violacea
C. cladosporioides	Pithomyces chartarum
C. oxysporum	Scolecobasidium selmum
Cochliobolus geniculatus	Stachybotrys atra
Epicoccum nigrum	Trichoderma viride
Fusarium oxysporum	Tripospermum spp.
Nigrospora spp.	Ulocladium atrum
Algae/cyanobacteria	
Calothrix spp.	Bacilliarophyta
Gloeocapsa spp.	Chlorella spp.
Leptolyngbya spp.	Chlorococcum spp.
Lyngbya spp.	Eustigmatos spp.
Michrochaete spp.	Klebshormidium spp.
Microcoleus spp.	Nannochloris spp.
Nostoc spp.	Stichococcus bariliaris
Oscillatoria spp.	Trebouxia spp.
Plectonema spp.	Trentepohlia aurea
Pleurocapsales	Trentepohlia odorata
Scytonema spp.	Tribonema spp.
Synechococcus spp.	Xanthophyta
Synechocystis spp.	

Specially formulated paints such as chlorinated rubber paints do not support microbial growth. They are used where there is a high risk of mould growth. Antifouling paints used on ships generally contain highly toxic metal-containing biocides which ensure protection against marine fouling organisms, often for as little as 2 years. The leaching efficiency of the metal in the biocide decreases with time, as the ions have farther to migrate through the paint film to the surface. Thus, although paint films which have failed may, on analysis, contain sufficient overall residual levels of the metal to ensure protection, the outer 50–75 μm of the film may itself be depleted, allowing the development of fouling. The self-polishing copolymers are sparingly soluble in water, so no depleted zone is formed. However, they are really of practical use only where there is a rapid flow of water over the surface, continually wearing away the coating to expose a fresh biocide-containing surface. They would not be suitable for static structures such as offshore oil rigs.

Other coatings, for example, epoxy resins, are usually nutritionally inert to microorganisms. Varnishes (paints without the pigment component) are an exception, in that they will support growth under favourable environmental conditions, resulting in staining and lifting of the film. This is often due to the wood substrate which, if not adequately preserved, allows the growth of fungi, which then pierce the varnish film. Renderings containing synthetic resins used as cement substitutes seem to be causing an increase in disfigurement problems. This may be due to the neutral pH of the render which allows algae and moulds to appear earlier than with traditional renderings.

Cosmetics and health products

Cosmetics are products which are designed to clean, beautify, or perfume the human body. They comprise four functional groups: skin, hair, oral care, and colour cosmetics, comprising creams, sunscreens, shampoos, styling gels, toothpaste, lipstick, mascara, and face powders. Many products contain water and a plethora of raw materials which are either very susceptible, such as anionic surfactants, or, because of their natural origins, are contaminated with bacteria and fungi from the environment in which they are sourced. Both the USFDA and the European Union Directive regarding the marketing of cosmetics state that they should be safe for use and not cause human health problems. After the presence of microorganisms in contaminated cosmetic products was first brought to the fore in 1965 in a Swedish report, various national surveys over the period 1967–78 have shown the extent and range of, in particular, pathogenic bacteria which are present in unused cosmetic preparations (Table 3.12). Some of these have been linked to infections, for example, eye infections from mascaras and hospital-acquired septicaemia from hand lotions and handcreams.

Not all problems associated with cosmetics are caused by pathogenic species. Shampoos and other anionic-surfactant-containing products are subject to the same effects as those which occur in liquid paint, that is, off-odours, discolouration, and loss in viscosity. The surfactant

Table 3.12. Gram-negative bacteria isolated from cosmetic products

Pseudomonas aeruginosa	Handcream
P. fluorescens	Hair oil
P. putida	Cleansing lotion
P. maltophila	Mascara
P. pseudoalcaligenes	Hair cream
Enterobacter cloacae	Toothpaste
Klebsiella aerogenes	Dental powder

content of shampoos is liable to degradation by bacteria, resulting in a slimy or ropy liquid with an unpleasant odour. An alkylsulfatase has been shown to be responsible for hydrolysis of the anionic surfactant sodium dodecyl sulfate used in shampoo formulations. Toilet soaps have been found to be colonized by *Aureobasidium pullulans, Stachybotrys atra, Scopulariopsis brevicaulis* and *Trichoderma viride.* Deteriorated facial and baby lotions have been found to contain species of *Serratia* and *Pseudomonas.* Creams and wet wipes may exhibit surface mould growth during storage (Figures 3.23 and 3.24.).

Natural earths, such as kaolin, talc, bentonite, and french chalk, can contain gram-negative and anaerobic-spore-bearing bacteria, as well as mould spores. Non-aqueous oily materials are less susceptible, but the ingress of water may result in the growth of moulds in, for example, liquid paraffin. Aqueous materials, such as surfactants, dyes, and water itself, are often the greatest source of contamination. Mains water invariably contains small numbers of microorganisms, and opportunist pathogens, such as *Pseudomonas,* can survive in minimal nutrient situations.

There are relatively few published accounts of the biodeterioration of pharmaceuticals. This is partly due to careful control measures exercised by manufacturers (Good Manufacturing Practice), the sensitive nature of the industry, and also the difficulties in elucidating the biodeterioration mechanisms. The presence of pathogens or the action of microorganisms capable of altering the structure of the active ingredient has been observed in pharmaceuticals. Thyroid and pancreas extracts have been reported to be contaminated with *Salmonella.* Hospital supplies of distilled water, destined for production of sterile fluids, have been found to contain *P. thomasii* and, in one particularly famous (or infamous) case

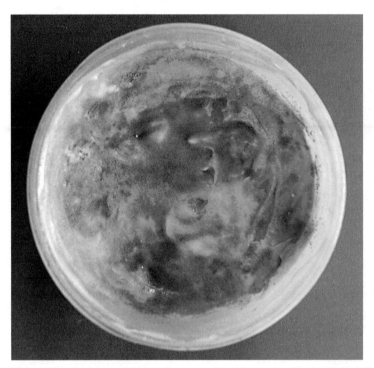

Figure 3.23. Surface mould growth on a moisturizing cream. Photo: Dr K. J. Seal.

in Brazil, toxin-producing cyanobacteria. *P. aeruginosa* has been isolated in significant numbers from pharmaceutical solutions containing cinnamon, peppermint, camphor, and aniseed.

Changes in activity of the active ingredient have been documented for antibiotics, aspirin, and other drugs. β-lactamases produced by a broad range of gram-positive and gram-negative bacteria can inactivate penicillins, chloramphenicol, streptomycin and kanamycin, and cephalosporins. Eyedrops containing atropine can lose up to 20% of their activity in the presence of *Corynebacterium* and *Pseudomonas* spp. *A. niger* and *Acinetobacter lwoffii* will degrade aspirin through catechol to *cis, cis*-muconate, whereas heroin can be degraded to morphine and strychnine by *A. lwoffii* and thalidomide can act as a carbon source for microorganisms in defined laboratory media. Some *Penicillium* species are able to utilize a series of substituted acetanilides (Table 3.13), such as paracetamol solution, to produce acetate and 4-aminophenol; similarly, phenacetin is degraded to acetate and 4-ethoxyaniline.

Figure 3.24. Surface mould growth on a wet wipe left; right-hand sample preserved. Photo: Dr K. J. Seal.

The utilization of amides by *A. nidulans* to produce the corresponding carboxylic acid and ammonia has led to postulation that drugs made from amide derivatives, such as the sulfonamides, pyrazinamide (antituberculosis), and niclosamide (a taenicide), may also be subject to degradation under adverse storage conditions.

Table 3.13. Acetanilides utilized by *Penicillium* sp. as a sole carbon source

4-aminoacetanilide	3-hydroxyacetanilide (metacetamol)
4-bromoacetanilide	4-hydroxyacetanilide (paracetamol)
2-carboxyacetanilide	4-hydroxy, N-methylacetanilide
4-carboxyacetanilide	4-iodoacetanilide
4-chloroacetanilide	4-methoxyacetanilide
2-ethoxyacetanilide	N-methylacetanilide
4-ethoxyacetanilide (phenacetin)	4-methylacetanilide
4-tormylacetanilide	4-nitroacetanilide
2- hydroxyacetanilide	

Bacteria and fungi may release toxic metabolites such as pyrogens into the product, or irritation or an allergic response may result from foreign proteins or outer membrane lipopolysaccharide components produced by the microorganism. Many of these toxins constitute the major microbial metabolites found in pharmaceutical products.

Metals

Corrosion of metals occurs when the structural elements undergo a chemical change from the ground state to an ionized species. It is an electrochemical phenomenon, consisting of an anodic (oxidation) reaction and an equivalent cathodic reaction, involving the reduction of a chemical species. The reactions for a ferrous metal in an aqueous oxygenated environment are

$$2Fe \longrightarrow Fe^{2+} + 4e, \qquad \textit{anodic reaction,} \qquad (3.1)$$
$$O_2 + 2H_2O + 4e \longrightarrow 4OH^-, \qquad \textit{cathodic reaction.} \qquad (3.2)$$

The iron goes into solution at the anode, leaving excess electrons which combine with oxygen to produce hydroxyl ions at nearby cathodic sites. In the absence of oxygen, either hydrogen ions or water is reduced at the cathode:

$$4H^+ + 4e \longrightarrow 2H_2, \qquad\qquad\qquad\qquad\qquad (3.3)$$
$$2H_2O + 4e \longrightarrow H_2 + 2OH^-. \qquad\qquad\qquad\qquad (3.4)$$

These reactions can be influenced by microbial activities, but the basic mechanisms of corrosion remain electrochemical. The resulting corrosion has been termed biocorrosion, microbially influenced corrosion (MIC), and biologically influenced corrosion, among others. There are no official figures for the cost of biocorrosion. Some idea can be gained from the following examples. In the 1950s, the costs of repair and replacement of pipelines damaged by MIC in the United States were estimated at around U.S.$0.5–2 billion per year. Replacement of biocorroded gas mains in the UK has been reported to cost £250 million per year. The South African power company, Escom, has detected biocorrosion of carbon steel in cooling water systems in almost all its power plants, and the costs associated with repairs and downtime are millions of dollars annually. It has been suggested that as much as 20% of all metal corrosion is influenced by microorganisms.

In the majority of cases, the process is associated with the formation of biofilms on the metal surface. The close association between the microbial cells and the material enables the metabolic activities to affect the metal surface directly. Because biofilms are never homogeneous, the resulting corrosion is typically pitting corrosion. Generalized corrosion is rarely caused by microorganisms, although this may occur when released metabolites, such as acids, are the responsible agents.

The complex reactions involved in biocorrosion explain why it has not been possible to relate microbial numbers to corrosion rates. Neither concentrations of cells in the planktonic (suspended) phase nor the adhered (sessile) numbers indicate the intensity or risk of corrosion, even when the microorganisms involved are known to be associated with corrosive activity. Not all microorganisms have this ability, and, indeed, some biofilms have been shown to be protective, although the mechanisms for this have not always been determined. Table 3.14 lists the types of microorganisms which may be associated with biocorrosion and the industrial environments in which they may be found.

It is important to realize that corrosion is rarely caused by a single factor or by a single type of microorganism. Abiotic factors are always involved, and these may enhance or retard the corrosion rate. The interactions between different microorganisms, which always occur in natural environments, may also result in increased or decreased corrosion. Controlled laboratory experiments have shown that the presence of the facultatively anaerobic bacterium, *Vibrio anguillarum*, reduced chemical corrosion of carbon steel, but increased that associated with the SRB, *Desulfovibrio vulgaris*. It was suggested that the tightly adherent biofilm produced by *V. anguillarum* protected the metal surface from chemical attack, but that in the presence of SRB the resultant mixed biofilm became superaggressive. Interestingly, the presence of a third facultative bacterium, itself without corrosive effect, modified the activity of the double culture. In other experiments, a mixed culture of a thermophilic *Bacillus* and the marine bacterium *Deleya marina* produced an extracellular polymer that reduced the corrosion rate of carbon steel by 94%.

Direct and indirect mechanisms of corrosion may be distinguished. Direct mechanisms occur where the organism is directly oxidizing or chelating the metal or metabolizing a component of the electrochemical process, such as hydrogen produced at the cathode. Indirect mechanisms include

Table 3.14. Microorganisms which may influence metal corrosion

Microorganisms	Main industrial areas where problems occur	Mechanism(s) of corrosion
Green algae and blue-green bacteria	Cooling towers and heat exchangers, metal structures in lakes and rivers	Concentration cells, hydrogenase activity
Filamentous fungi	As above, fuel storage tanks, MWFs	Concentration cells, acid production
Aerobic heterotrophic bacteria	As above	As above
Sulfur-oxidizing bacteria	Waste treatment plants, metal mining	Production of sulfuric acid
Iron-oxidizing bacteria	Oxygenated water systems, injection wells in the petroleum industry	Direct metal oxidation, concentration cells
Metal-depositing bacteria	As above	Concentration cells
Metal-reducing bacteria	Anaerobic areas in soils and below deposits	Destabilization of protective (passive) films
SRB	As above, secondary recovery systems in the oil industry, liquid storage tanks, pulp and paper industry, heat exchangers, MWFs	Various (see text)
Slime-producing microorganisms	Any area where biofilms form	Concentration cells, chelation of metal ions

microorganisms setting up differential concentration (electrochemical) cells on metal surfaces, destabilizing passive films, or producing acidic metabolites, all of which lead to corrosion. Microorganisms might use the following mechanisms in initiating or accelerating corrosion processes:

1. setting up of concentration cells
2. release of metabolic products
3. removal of cathodic hydrogen.

As will be seen, however, it is rare to find a single mechanism operating in isolation.

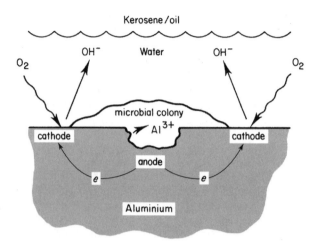

Figure 3.25. Electrochemical cell set up beneath a microbial colony on a metal surface.

MICROBIAL CONCENTRATION CELLS

Microbial concentration cells are often referred to as *differential aeration cells*, because they most frequently arise from an oxygen gradient resulting from depletion of oxygen by the microbial colony where it is in contact with the metal. The edge of the colony becomes cathodic where oxygen is present, whereas the centre is anodic and metal ions are released (Figure 3.25); these may combine with anions present at the site to form an insoluble deposit on the surface which may, if the corrosion product deposit is of a suitable composition, inhibit further corrosion.

The flow of electrons from anode to cathode produces a corrosion current, which can be measured. Figure 3.26 shows a diagram of the potentiodynamic corrosion curve, produced by the determination of the current produced in a metal on the application of a varying potential, measured with respect to a standard electrode such as the saturated calomel electrode (SCE). This type of electrochemical analysis is frequently used to determine the corrosive activity of an environment, either with or without living microorganisms and/or their metabolic products. The pitting potential, Ep, is that at which the passive layer on the surface of the metal is ruptured and pitting begins to occur. The lower the potential at which this happens, the more likely is the metal to corrode, or, in other words, the more corrosive is the environment. Many other techniques, some of them also electrochemical, are used in the study of biocorrosion, but it is

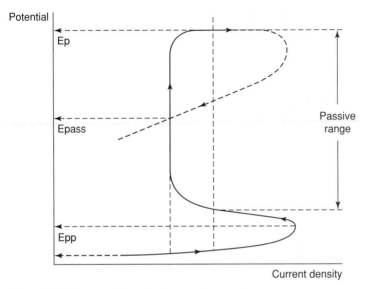

Figure 3.26. Diagram of a potentiodynamic corrosion curve.

not the place of this book to cover them and those readers interested are advised to consult the specialist literature.

Potential differences of 60 mV have been recorded across colonies of microorganisms on metal surfaces. Even if the centre of the colony dies, the electrochemical cell can remain active, oxygen being consumed during autolysis and oxidation of the dead tissue. A similar phenomenon occurs even in the absence of microbial cells. For example, artificial biofilms formed of polysaccharides (xanthan, alginate, and agarose) were shown to have a considerable influence on the corrosion of copper. However, it is difficult to differentiate the concentration cell mechanism from the chelation of metal ions, which also occurs in the presence of these polymers. Direct involvement of extracellular polymeric substances (EPSs) produced by living cells in the biodeterioration of stainless steel, copper, and carbon steel has been demonstrated, and it is likely that both these mechanisms (and possibly others) are involved. Figure 3.27 is a scanning electron micrograph of a SRB biofilm on a stainless steel surface. The EPS [contracted into strands by the preparation procedures for scanning electron microscopy (SEM)] is clearly seen.

The iron bacteria (*Sphaerotilus, Crenothrix, Leptothrix*, and *Gallionella*) are chemolithotrophs, gaining energy from the oxidation of ferrous to ferric ions. In doing this, they deposit ferric hydroxide on the surface

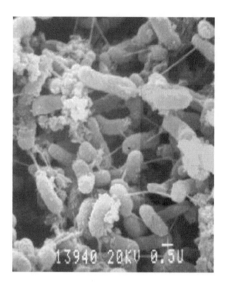

Figure 3.27. Biofilm formed by SRB on a stainless steel surface. Photo: Dr Christine Gaylarde.

of iron to which they may be attached (e.g., pipework). This produces tubercles beneath which corrosion can occur. Some bacteria in the preceding groups can also oxidize manganous to manganic ions, depositing manganese dioxide on the metal surface and destabilizing passive oxide films, as well as contributing to tubercle formation. Concentration cell corrosion can occur in water recirculating systems, in which increases in salt concentrations and organic loadings can enable bacteria, fungi, and algae to form slime on metal surfaces. The perforation of aluminium alloy fuel tanks on jet aircraft can be due to bacterial or fungal colonies growing as a biofilm on the metal surface, resulting in concentration cell corrosion or in acid corrosion by localized metabolite production.

METABOLIC PRODUCT SECRETION

Microorganisms secrete a range of acidic and other metabolites into their immediate environment. Organic acids, hydrogen sulfide, and ammonia may accumulate. The sulfur-oxidizing bacteria produce sulfuric acid, which is highly corrosive both to metals and concrete products. The fungus *Hormoconis resinae* causes corrosion of aircraft aluminium alloys by the secretion of citric, iso-citric, *cis*-aconitic, and α-oxoglutaric acids, resulting in pitting and selective removal of zinc, magnesium, and aluminium, leaving copper and iron aggregates. This suggests that organic acid corrosion plays a more important role in pitting corrosion than the differential aeration cell. Some sulfur-oxidizing bacteria can oxidize elemental sulfur and its reduced forms to obtain energy, thus producing the corrosive metabolite, sulfuric acid. *Acidithiobacillus ferrooxidans* is also able to oxidize ferrous ions, hence causing corrosion by a direct mechanism. This bacterium is important in the leaching of metal ores. In leaching iron from iron pyrites, the overall reaction is

$$4FeS_2 + 15O_2 + 2H_2O \longrightarrow 2Fe_2(SO_4)_3 + 2H_2SO_4. \qquad (3.5)$$

Both the iron and the sulfur are oxidized in this reaction. *A. thiooxidans* can grow only at a pH below 5 and is capable of producing a highly corrosive environment with a pH of 0.7, equivalent to about 5% sulfuric acid. A number of interlinked reactions have been put forward to explain the formation of sulfuric acid by mixed cultures of thiobacilli:

$$2H_2S + 2O_2 \longrightarrow H_2S_2O_3 + H_2O, \qquad (3.6)$$

$$5S_2O_3^{2-} + 4O_2 + H_2O \longrightarrow 5SO_4^{2-} + H_2SO_4 + 4S^0, \qquad (3.7)$$

$$2S^0 + 3O_2 + 2H_2O \longrightarrow 2H_2SO_4. \qquad (3.8)$$

The sulfur-containing substrates originate from either the activities of the SRB or fermentative breakdown of sulfur-containing proteins, producing sulfides in the same locality as the sulfur oxidizers (this ecosystem is known as a sulfuretum). Sulfureta are found in sewage pipes, compacted reclaimed land, or sediments in riverbeds. Sulfate-containing waters are a source of energy for the SRB. Acid mine waters, which cause corrosion of mining equipment in pits, are the result of *A. ferrooxidans* in contact with iron ore. The pH drops to between 2 and 3, and deposits of ferric sulfate are formed. Severe corrosion of concrete and iron sewage pipes can occur at the air/sewage interface when flow rates are lower than the design has allowed. Lack of flushing through results in reduced aeration and the build-up of sulfide, which is then oxidized to sulfuric acid.

Some bacteria, such as species of *Shewanella* and *Pseudomonas*, can reduce iron and/or manganese oxides. This can result in the protective (passive) oxide layers on metal surfaces being lost or replaced with less stable (reduced) films. This is a mechanism of corrosion which has been little studied so far.

Sulfate-reducing bacteria

The SRB are a heterogeneous group of anaerobic bacteria whose only common characteristic is the use of sulfates as electron acceptors in the

Figure 3.28. Formation of iron sulfide on iron nails in a 72-h culture of (a) *Desulfovibrio vulgaris* grown on Postgate's medium C; (b) a fresh culture less than 24-h old. Photo: Dr K. J. Seal.

respiratory chain. This sulfate respiration is also known as dissimilatory sulfate reduction and its final product, apart from energy, is sulfide [see reaction (3.9)]. A large number of bacteria with this activity have been identified, in addition to the two genera initially described (*Desulfovibrio* and *Desulfotomaculum*). They belong to the phyla Thermodesulfobacteria and Nitrospirae and the classes Deltaproteobacteria and Clostridia of the domain Bacteria. Some members of the domain Archaea can reduce other oxidized forms of sulfur, but utilization of sulfate appears to be restricted to the Bacteria.

$$SO_4^{2-} + 4H_2 \longrightarrow S^{2-} + 4H_2O. \tag{3.9}$$

To carry out this reaction, SRB require not only anaerobic conditions, but a reduced environment with a redox potential of -150 mV (normal hydrogen electrode). The production of sulfide encourages the maintenance of the low potential and enhances the possibility of corrosion as a result of the precipitation of iron sulfide [see reaction (3.14) and Figure 3.28].

In 1934, von Wolzogen Kühr and van der Vlugt published their "cathodic depolarization theory" of the mechanism of biocorrosion by SRB. This is shown in the following reactions:

$$4Fe \longrightarrow 4Fe^{2+} + 8e, \qquad \textit{anodic reaction,} \qquad (3.10)$$

$$8H_2O \longrightarrow 8OH^- + 8H^+, \qquad\qquad (3.11)$$

$$8H^+ + 8e \longrightarrow 4H_2, \qquad \textit{cathodic reaction,} \qquad (3.12)$$

$$SO_4^{2-} + 4H_2 \longrightarrow S^{2-} + 4H_2O, \qquad \textit{cathodic depolarization,} \qquad (3.13)$$

$$Fe^{2+} + S^{2-} \longrightarrow FeS, \qquad \textit{corrosion product,} \qquad (3.14)$$

$$3Fe^{2+} + 6OH^- \longrightarrow 3Fe(OH)_2, \qquad \textit{corrosion product.} \qquad (3.15)$$

The overall reaction is

$$4Fe + SO_4^{2-} \longrightarrow FeS + 3Fe(OH)_2 + 2OH^-. \qquad (3.16)$$

Reaction (3.13) represents the role SRB may play in enhancing corrosion. Removal of the hydrogen produced in the cathodic reaction [reaction (3.12)] by SRB activity allows the anodic reaction [reaction (3.10), in which iron goes into solution] to continue. Reactions (3.12) and (3.13) are catalysed by the enzymes hydrogenase and sulfate reductase (a series of enzymes), and one of the final products is corrosive sulfide, produced through a number of still incompletely defined steps. However, tests on strains of SRB which lack hydrogenase showed that corrosion rates remain high, and an alternative mechanism was put forward to explain this phenomenon.

Ferrous sulfide formed as a consequence of the reduction of sulfate to sulfide by SRB is corrosive towards mild steel and plays an important role in the corrosion process under anaerobic conditions. Even after formation of these sulfides, however, bacterial activity remains necessary for continuing corrosion. As the iron sulfide becomes saturated with hydrogen atoms, polarization occurs, reducing the corrosion rate. Cell activity depolarizes the deposits, allowing further corrosion. It has also been shown that biogenically produced sulfides are more aggressive than chemically formed sulfides. A number of other mechanisms have been suggested over the years to explain the action of SRB on metals, and these are shown in Table 3.15. That the mechanism of corrosion by SRB is different from that of less aggressive, acid-producing bacteria is indicated by the morphology of the corrosion attack. Figure 3.29(a) shows the generalized pitting caused on carbon steel by a member of the genus *Vibrio*. This should be compared with the characteristic attack by the SRB *D. vulgaris*, shown in Figure 3.29(b), in which metal grains have been removed and raised grain boundaries containing multiple small pits can be seen.

Table 3.15. Suggested corrosion mechanisms of SRB

Mechanism	Reference
Cathodic depolarization (hydrogenase enzyme)	von Wolzogen Kühr and van der Vlugt, 1934
Anodic depolarization	Salvarezza and Videla, 1986
Sulfides	King and Wakerley, 1973
Elemental sulfur from sulfide oxidation	Hardy and Brown, 1984
Fragilization caused by H_2	Edyvean et al., 1988
Volatile P compound	Iverson and Ohlson, 1983
Metal binding by EPS	Beech and Cheung, 1995

SRB are widespread, having been isolated from soils throughout the world and in all kinds of waters from fresh to saline. They also occur, as would be expected, in anaerobic clays and estuarine muds and have even been isolated from the human gut. They have been said to be present in oil-bearing strata; whether they originated there or came by means of seepage is difficult to demonstrate. SRB have been implicated in the corrosion of fuel tanks, metal pipes, and other ironwork buried in the soil or in estuaries, and estimates have suggested that around 50% of pipework failures were due to bacterial corrosion of the sulfide type. Figure 3.30 shows a steel pipe which was perforated after only 2 years in the ground by SRB activity. Some SRB survive, but are not active, in aerobic conditions for example, highly aerated liquids can harbour SRB below biofilms and deposits. Indeed, it is below biofilms that these bacteria may be at their most active, receiving at these sites considerable help for their rather fastidious growth from the other organisms present in the consortium.

Cast iron suffers from graphitization, in which the iron is solubilized, leaving a network of graphite which is easily penetrated. Large concentrations of SRB can also occur in metal-working emulsions and large recirculating water systems, in which they can cause equipment failure and corrosion of metal products. The offshore oil industry has long been concerned over the souring of oil which is recovered from oil-winning operations. Sour oil contains hydrogen sulfide, which is microbiologically generated by SRB. Other problems caused by SRB in the oil industry include corrosion of pipelines, pumps and deaeration towers from which oxygen is routinely removed to prevent corrosion. The major source of the contamination is the seawater, which is injected into the oil reservoirs daily in large amounts

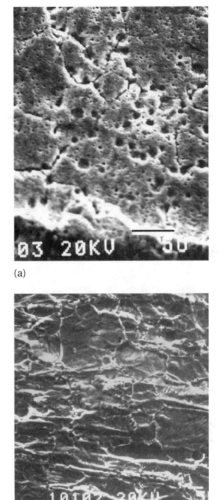

Figure 3.29. Scanning electron micrographs of carbon steel corroded by contact with pure cultures of bacteria: (a) generalized pitting produced by metabolic products of *Vibrio* cells, (b) removal of grains and pitted grain boundaries caused by SRB. Photo: Dr Christine Gaylarde.

(a)

(b)

(600 000 barrels/day) to maintain the pressure in the well (secondary recovery). Sulfide also inhibits the formation of molecular hydrogen at the cathode [see reaction (3.11)], allowing atomic hydrogen to diffuse into the steel. If the hydrogen diffuses into inclusions in the steel, produced during the rolling process, it can form molecular hydrogen (having a greater mass than two atoms of hydrogen), resulting in internal stresses and an effect known as hydrogen cracking. This can cause failures in steel pipelines

Figure 3.30. Steel pipe corroded through within 2 years of being placed in an anaerobic soil. Photo: Dr Christine Gaylarde.

containing high levels of hydrogen sulfide and is thus of concern to the offshore oil industry.

The use of cheap biocides such as hypochlorite and glutaraldehyde in injection waters and drilling muds has been widespread, with limited effectiveness, particularly in the control of biofilms on the internal surfaces of pipelines and storage vessels. Benzothiazololes, dithiocarbamates, bisthiocyanates, and isothiazolinones are some of the more expensive, but more effective, options.

The involvement of microorganisms in the corrosion of external structures at intertidal and splash zones is less studied, but has been shown to be an important contribution to increased perforation rates of steel piling in ports and harbours. The fouling communities of algae, anemones, and barnacles which attach themselves to the platform legs have been shown to harbour SRB deep within the biofilms.

In conclusion, the process of microbial corrosion takes a number of forms, some better documented than others. The difficulty arises in establishing the relative importance of corrosion by microorganisms over other types of corrosion in the same environment and the interrelationships between the biological influences and conventional corrosion mechanisms. Detailed studies, involving advanced surface science techniques and microbiological activity analyses on a microscale (see Chapter 5) are essential for determining the real importance of microorganisms in natural metal corrosion.

Figure 3.31. Surface growth on an adhesive. Photo: Dr K. J. Seal.

Adhesives and sealants

Despite little published information specific to this area, we may approach the subject by examining the components of the products to assess their susceptibility to microorganisms. However, in practice, biocides are often used to protect susceptible formulations and there is some field experience upon which to call. Traditional adhesives and sealants were derived from natural materials such as proteins (casein), carbohydrates (starch), and petroleum (bitumen). All of these materials are susceptible to deterioration in varying degrees. More recently plastic resins and synthetic rubber-like materials have been increasingly employed because of their improved in-service and fitness-for-use properties.

   Examples of water-based adhesives are those which contain casein, starch, polyvinyl acetate, methyl cellulose, and latex-emulsion-based systems. This type of adhesive is more likely to give problems than the solvent-based type, particularly during storage (see Figure 3.31). Considering only the main constituent, it is clear that spoilage can occur.

Table 3.16. Organisms associated with adhesive failures

Bacteria	Yeasts	Fungi
Aerobacter aerogenes	Saccharomyces cerevisiae	A. niger
Bacillus mycoides	Rhodotorula sp.	A. pullulans
Escherichia coli		Chaetomium globosum
Proteus mirabilis		Geotrichum sp.
P. aeruginosa		P. luteum
		S. brevicaulis
		T. viride

Table 3.16 lists organisms which have been found to be associated with adhesive failures.

Even in a seemingly susceptible system, the pH may be highly alkaline or acidic, restricting growth. However, experience has shown that this is not necessarily an effective control method. The solvent-based products are significantly less prone to storage problems, but can be affected as a cured film when in service.

Epoxy resins have excellent resistance, whereas polyesters and polyurethanes, depending on their structure, are variously susceptible. A range of formaldehyde-containing resins have been tested for their resistance to the growth of wood-destroying fungi. These resins are used as glues for particleboard manufacture, and certain types encourage fungal growth on the boards. The most susceptible types have been found to be based on urea-formaldehyde and phenol-formaldehyde. The inclusion of melamine in the urea type reduced utilization, whereas a combination of phenol and resorcinol strongly inhibited the test fungi.

Sealants are made from a range of materials which are capable of adhesion and flexibility and have the ability to contain or exclude water. Rubbers and bitumens have been used for many years, and more recently polyurethanes, epoxides, and polysulfides have been introduced. They are commonly used in sewage installations as joints between concrete tank sections. Breakdown of the sealant and the consequent leakage of fluid has been attributed to the activities of microorganisms. An examination of a range of sealants, covering the main chemical types previously listed, showed that they were subject to biodeterioration within 3-month immersion at a range of test sites around a sewage disposal plant. At the time of the test, only the neoprene- and epoxy-based products showed any resistance. Polysulfide sealants have now been improved by alteration of the curing mechanism.

Magnetic media – information technology

Audio, audio/visual, and computer media play an increasingly important role in our daily lives as a means of transmitting, exchanging, and storing information. Most of the software relating to information technology is based on plastics, generally polyesters, polyurethanes, and polycarbonates. As we have seen, some of these may be susceptible to biodeterioration. The packaging material is also important; the cellulosic fibre lining in floppy disks and the plastic cases of videotapes and CDs can provide extra nutrients in either solid or gaseous form. Moisture, present in humid air, is the only additional requirement.

Floppy disks and videotapes

Videotape and floppy disks, or diskettes, consist of magnetic particles, usually ferric oxide, bonded to a polyester base; the magnetic particles are commonly applied in a polyurethane slurry. Other ingredients are also present, such as plasticizers, abrasives, lubricants and dispersants. All of these may be susceptible to microbial growth. Species of the fungal genera *Aspergillus, Penicillium*, and *Paecilomyces* have been isolated from, and shown to grow on the surface of, floppy disks. These are not plastic-degrading organisms, and, indeed, degradation of the plastic core is not necessary for the thin magnetic surface layer to become sufficiently contaminated to interrupt electrical conductivity, leading to failure. One of the authors (CCG) has isolated the cellulose-degrading fungus *Chaetomium*, from unusable floppy disks. Presumably this was feeding on the fibre inserts between the magnetic medium and the plastic case. Attempting to run a floppy disk with particulate matter on the surface can permanently destroy data on the disk and damage the computer. The media should be cleaned and the information transferred to a new floppy disk.

Videotape carries a combination of video, audio, and transport speed control information. When played, the tape must pass over various posts, guides, and heads to obtain good results. The tolerance between tape and playing head is less than one-quarter the thickness of a human hair; only a small fungal colony is required to interfere with efficient reading (see Figure 3.32). Even worse, playing a damaged or contaminated tape on a VCR can ruin the tape and draw out the lubricant from the drive mechanism of the VCR, causing serious damage to the equipment. Specialized firms exist for the cleaning and recuperation of disks and tapes, and these may be found through the Internet.

Figure 3.32. Mould growth on videotape. Photo: Dr D. Allsopp.

Compact disks

A compact disk or CD-ROM comprises three or four layers:

1. The plastic disk: a thick layer of polycarbonate plastic which provides protection, flexibility and shape. The digital information is stamped onto the top surface of the disk as a series of grooves.
2. Reflective foil: a thin layer of aluminum or gold.
3. Lacquer layer.
4. Graphic layer: a thin layer identifying the digitalized contents of a CD-ROM, which may or may not be present.

In the case of recordable optical disks, an additional dye layer is present.

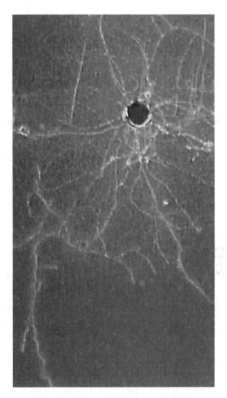

Figure 3.33. Fungal growth on the surface of a CD. Photo: Dr Christine Gaylarde.

There have been few reports describing biodegradation of polycarbonates, which are chemically very stable. However, growth of fungi on CDs is a common occurrence in hot and humid environments (see Figure 3.33). In 2001, a fungus resembling *Geotrichum candidum* was reported growing on a music CD from Belize, and the Spanish investigators showed a scanning electron micrograph indicating degradation of the surface. They stated that the aluminium layer was completely destroyed. Whether the polycarbonate was degraded seems to be doubtful. The report engendered a lot of public interest and comments on the Internet. Phillips, the inventor of the CD, issued a statement that this was a "one-off" case, possible only under the special environmental conditions of Belize!

Simple cleaning with soap and water and a soft cloth will remove microbial growth, although etching may already have occurred. In this case, a fine water-based metal polish with no ammonia, such as that used for cleaning silver, may be used to polish the surface with a good-quality fine cotton cloth. This treatment is not always successful! Once again, specialized firms may be consulted.

REFERENCES AND SUGGESTED READING

Magnetic media

Bardage, S. L. and Daniel, G. (1997). The ability of fungi to penetrate micropores: implications for wood surface coatings. *Mater. Org.*, **31**, 233–45.

Garcia-Guinea, J., Cárdenes, V., Martínez, A. T., and Martínez, M. J. (2001). Fungal bioturbation paths in a compact disk. *Naturwissenschaften*, **88**, 351–4. Also published online as Short Communication DOI: 10.1007/s001140100249

McCain, J. W. and Mirocha, J. (1994). Screening computer diskettes and other magnetic media for susceptibility to fungal colonization. *Int. Biodeterior. Biodeg.*, **33**, 255–68.

Suyama, T., Hosoya, H., and Tokiwa, Y. (1998). Bacterial isolates degrading aliphatic polycarbonates. *FEMS Microbiol. Lett.*, **161**, 255–61.

Fuels and lubricants

Bailey, C. A. and May, M. E. (1979). Evaluation of microbiological test kits for hydrocarbon fuel systems. *Appl. Environ. Microbiol.*, **37**, 871–7.

Batts, B. D. and Fathomi, A. Z. (1991). A literature review on fuel stability studies with particular emphasis on diesel oil. *Energy Fuels*, **5**, 2–21.

Bennett, E. O. (1974). Water quality and coolant life. *Lubr. Eng.*, **30**, 549–55.

Bento, F. M. and Gaylarde, C. C. (2001). Biodeterioration of stored diesel oil: Studies in Brazil. *Int. Biodeterior. Biodeg.*, **47**, 107–12.

Byers, J. P. (Ed.). (1994). *Metalworking Fluids*. Marcel Dekker, New York.

Cafone, L. Jr., Walker, J. D., and Cooney, J. J. (1973). Utilization of hydrocarbons by *Cladosporium resinae. J. Gen. Microbiol.*, **76**, 243–6.

Cook, P. E. and Gaylarde, C. C. (1988). Biofilm formation in aqueous metal working fluids. *Int. Biodeterior.*, **24**, 265–70.

Coughlin, R. W., Williams, D., Seveau, E., Veith, R., and Howes, T. D. (1992). Enumeration of microorganisms in metalworking fluids using photometric methods. *CIRP Ann. Int. Inst. Prod. Eng.*, **41**, 357–60.

Davis, J. B. (1967). *Petroleum Microbiology*. Elsevier, Amsterdam.

Elsmore, R. (1989). Survival of *Legionella pneumophila* in dilute metalworking fluids. *Tribol. Int.*, **22**, 213–17.

Gaylarde, C. C., Bento, F. M., and Kelley, J. (1999). Microbial contamination of stored hydrocarbon fuels and its control. *Rev. Microbiol.*, **30**, 1–10.

Haggatt, R. D. and Morchat, R. M. (1992). Microbiological contamination: Biocide treatment in naval distillate fuel. *Int. Biodeterior. Biodeg.*, **25**, 87–99.

Hartman, J., Geva, J., and Fass, R. (1991). A computerized expert system for diagnosis and control of microbial contamination in jet fuel and diesel fuel storage systems. In *Proceedings of the Fourth International Conference on Stability and Handling of Liquid Fuels*, pp. 153–66.

Hendey, N. I. (1964). Some observations on *Cladosporium resinae* as fuel contaminant and its possible role in the corrosion of aluminium alloy fuel tanks. *Trans. Br. Mycol. Soc.*, **47**, 467–75.

Herwaldt, L. A., Gorman, G. W., McGrath, T., Toma, S., Brake, B. and Hightower, A. W. (1984). A new *Legionella* species, *Legionella feeleii* species Nova, causes Pontiac fever in an automobile plant. *Ann. Intern. Med.*, **100**, 333–8.

Hettige, G. E. G. and Sheridan, J. E. (1989). Interactions of fungi contaminating diesel fuel oil. *Int. Biodeterior.*, **25**, 299–316.

Hettige, G. E. G. and Sheridan, J. E. (1994). The mycoflora in aviation and diesel fuel. In *Recent Advances in Biodeterioration and Biodegradation*, Garg, K.L., Garg, N., and Mukerji, K.G. (Eds.). Naya Prokash, Calcutta, Vol. II, pp. 17–29.

Hill, E. C. and Al-Zubaidy, T. (1979). Some health aspects of infections in oil and emulsions. *Tribol. Int.*, **8**, 161–4.

Hill, E. C. (1993). Microbial aspects of health hazards from water based metal working fluids. *Tribol. Int.*, **16**, 136–40.

Hill, E. C. (1987). Fuels. In *Microbial Problems in the Offshore Oil Industry*, Hill, E.C., Shennan, J., and Watkinson, R. (Eds.). Wiley, New York, pp. 219–30.

Hill, E. C. and Hill, G. C. (1993). Microbiological problems in distillate fuels. *Trans. Inst. Mar. Eng.*, **104**, 119–30.

Passman, F. J. (1992). Controlling microbial contamination in metal working fluids. Presented at the conference on metal working fluids, March 16–18, 1992, Cincinnati, OH.

Passman, F. J. (1995). Biocide toxicity: A comparison of the toxicological properties of common metalworking fluid biocides. In *Symposium Proceedings on the Industrial Metalworking Environment: Assessment and Control*. American Automobile Manufacturers Association, Detroit, MI, pp. 82–7.

Rinkus, K. M., Lin, W., Jha, A., and Reed, B. E. (1997). Investigation into the nature and extent of microbial contamination present in a commercial metalworking fluid. In *Proceedings of the Industrial Waste Conference*, pp. 601–10.

Rossmore, H. W. (1974). Microbiological causes of cutting fluid deterioration. SME Technical Paper, MR74–169.

Rossmoore, H. W. (1975). Extending cutting fluid life. *Manuf. Eng.*, **75**(5), pp. 27–8.

Shennan, J. L. (1988). Control of microbial contamination of fuels in storage. In *Biodeterioration 7*, Houghton, D.R., Smith, R.N., and Eggins, H.O.W. (Eds.). Elsevier, Barking, England, pp. 248–54.

Skoeld, R. O. (1991). Field testing of a model waterbased metalworking fluid designed for continuous recycling using ultrafiltration, *Lubr. Eng.*, **47**, 653–9.

Smith, R. N. (1987). Fuel testing. In *Microbiology of Fuels*, Smith, R.N. (Ed.). Institute of Petroleum, London, pp. 49–54.

Smith, R. N. (1991). Developments in fuel microbiology. In *Biodeterioration and Biodegradation 8*, Rossmoore, H. (Ed.). Elsevier, London, pp. 112–24.

Solana, M. J. V. and Gaylarde, C. C. (1995). Fungal contamination and growth in various hydrocarbon fuels. In *Biodeterioration and Biodegradation 9*, Bousher, E., Chandra, M., and Edyvean, R. (Eds.). Institute of Chemical Engineers, Rugby, England, pp. 621–8.

Wycislik, E. T. and Allsopp, D. (1983). Heat control of microbial colonisation of shipboard fuel systems. In *Biodeterioration 5*, Oxley, T. A. and Barry, S. J. Wiley, (Eds.). London, pp. 453–61.

Metals

Beech, I. B. and Cheung, C. W. S. (1995). Interactions of exopolymers produced by sulphate-reducing bacteria with metal ions. *Int. Biodeterior. Biodeg.*, **35**, 59–72.

Beech, I. B. and Gaylarde, C. C. (1999). Recent advances in the study of biocorrosion – an overview. *Rev. Microbiol.* **30**, 177–90.

Edyvean, R. G. J., Benson, J., Thomas, C. J., Beech, I. B., and Videla, H. A. (1988). Biological influences on hydrogen effects in steel in seawater. *Mater. Perform.*, **37**, 40–4.

Guiamet, P. and Gaylarde, C. C. (1996). Activity of an isothiazolone biocide against *Hormoconis resinae* in pure and mixed biofilms. *World J. Microbiol. Biotechnol.*, **12**, 395–7.

Hardy, J. A. and Bown, J. L. (1984). Sulphate-reducing bacteria: their contribution to the corrosion process. *Corrosion*, **40**, 650–54.

Iverson, I. P. and Ohlson, G. J. (1983). Anaerobic corrosion by sulphate-reducing bacteria due to highly reactive volatile phosphorus compound. In *Microbial Corrosion*, London, Metals Society, pp. 46–53.

King, R. A. and Wakerley, D. S. (1973). Corrosion of mild steel by ferrous sulphide. *Br. Corros. J.*, **8**, 41–54.

Parker, L. H. J., Seal, K. J., and Robinson, M. J. (1988). Hydrogen adsorption during the microbial corrosion of steel. In *Biodeterioration 7*, Houghton, D. R., Smith, R. N, and Eggins, H. O. W. (Eds.). Elsevier Applied Science, London, pp. 391–7.

Postgate, J. R. (1984). *The Sulphate-Reducing Bacteria*, 2nd ed. Cambridge University Press, Cambridge, England.

Salvarezza, R. C. and Videla, H. A. (1986). Electrochemical behaviour of aluminium in *Cladosporium resinae* cultures. In *Biodeterioration 6*, Barry, S. and Houghton, D. R. (Eds.). Elsevier, London, pp. 212–17.

von Wolzogen Kühr, C. A. H. and van der Vlugt, L. S. (1934). De grafiteering van Gietijzer als electrobiochemisch Proces in anaerobe Grunden. *Water (den Haag)*, **18**, 147–51.

Plastics and rubbers

Fu, M. H. and Alexander, M. (1992). Biodegradation of styrene in samples of natural environments. *Environ. Sci. Technol.*, **26**, 1540–4.

Howard, G. T. (2002). Biodegradation of polyurethane: A review. *Int. Biodeterior. Biodeg.*, **49**, 245–52.

Ishigaki, T., Sugano, W., Ike, M., and Fugita, M. (2000). Enzymatic degradation of cellulose acetate plastic by novel degrading bacterium *Bacillus spp* S2055. *J. Biosci. Bioeng.*, **90**, 400–6.

Keursten, G. T. G. and Groenevelt, P. H. (1988). Biodegradation of rubber particles in soil. In *Biodeterioration 7*, Houghton, D. R., Smith, R. N., and Eggins, H. O. W. (Eds.). Elsevier Applied Science, London, pp. 329–33.

Mukai, K. and Yoshiharu, D., (1995). Microbial degradation of polyesters. *Prog. in Ind. Microbiol.*, **32**, 189–204.

Oda, Y. et al. (1997). Polycaprolactone depolymerase produced by *Alcaligenes faecalis. FEMS Microbiol. Lett.*, **152**, 339–43.

Pranamuda, H., Chollakup, R., and Tokiwa, Y. (1999). Degradation of polycarbonate by a polyester-degrading strain, *Amycolatopsis sp.* Strain HT-6. *Appl. Environ. Microbiol.*, **65**, 4220–2.

Shuttleworth, W. A. and Seal, K. J. (1986). A rapid technique for evaluating the biodeterioration potential of polyurethane elastomers. *Appl. Microbiol. Biotechnol.*, **23**, 407–9.

Seal, K. J. (1988). The biodeterioration and biodegradation of naturally occurring and synthetic plastic polymers. *Biodeterior. Abstr.*, **2**, 295–317.

Trejo, A. G. (1988). Fungal degradation of polyvinyl acetate. *Ecotoxicol. Environ. Safe.*, **16**, 25–35.

Tsuchii, A., Takeda, K., and Tokiwa, Y. (1997). Degradation of the rubber in truck tires by a strain of *Nocardia. Biodegradation.*, **7**, 405–13.

Vert, M., Feijen, J., Albertsson, A., Scott, G., and Chiellini, J. (Eds.). (1992). Biodegradable polymers and plastics. *R. Soci. Chem. Spec. Publ. No. 109.*

Glass

Gorbushina, A. A. and Palinska, K. A. (1999). Biodeteriorative processes on glass: Experimental proof of the role of fungi and cyanobacteria. *Aerobiologia*, **15**, 183–91.

Kerner-Gang, W. (1977). Evaluation techniques for resistance of optical lenses to fungal attack. In *Biodeterioration Investigation Techniques*, A. H. Walters (Ed.). London, pp. 105–14.

Krumbein, W. E., Urzi, C. E., and Gehrmann, C. (1991). Biocorrosion and biodeterioration of antique and medieval glass. *Geomicrobiol. J.*, **9**, 139–60.

Perez-y-Yorba, M., Dallas, J. P., Bauer, C., Bahezre, C., and Martin, J. C. (1980). Deterioration of stained glass by atmospheric corrosion and micro-organisms. *J. Mater. Sci.*, **15**, 1640–7.

Thorset, I. H., Furnes, H., and Heldal, M. (1992). The importance of microbiological activity in the alteration of natural basaltic glass. *Geochim. Cosmochim. Acta*, **56**, 845–50.

Paints, latices, and adhesives

Cresswell, M. A. and Holland, K. (1995). Preservation of aqueous-based synthetic polymer emulsion and adhesive formulations. In *Preservation of Surfactant Formulations*, Morpeth, F. F. (Ed.). Chapman & Hall, Glasgow, pp. 212–61.

Gaylarde, P. and Gaylarde, C. C. (1999). Algae and cyanabacteria on painted surfaces in Southern Brazil. *Rev. Microbiol.*, **30**, 209–13.

Gillat, J. W. (1991). The microbiological spoilage of emulsion paints during manufacture and its prevention. *J. Oil Colour Chem. Assoc.*, **74**, 324–8.

Gillat, J. W. (1991). The need for antifungal and antialgal additives in high performance surface coatings. *J. Oil Colour Chem. Assoc.*, **74**, 197–203.

Gillat, J. W. (1995). Evaluating biocidal efficacy in polymer emulsions. Part 1, Establishment of a recommended microbial inoculum. *Paint Ink Int.*, **8**, 18–26.

Shirakawa, M. A., Gaylarde, C. C., Gaylarde, P. M., John, V., and Gambale, W. (2002). Fungal colonization and succession on newly painted buildings and the effect of biocide. *FEMS Microbiol. Ecol.*, **39**, 165–73.

Tothill, I. E., Best, D. J., and Seal, K. J. (1993). Detection of cellulolytic enzymes in water borne paint. *Int. Biodeterior. Biodeg.*, **32**, 115–28.

Tothill, I. E., Best, D. J., and Seal, K. J. (1993). Studies on the inhibitory effect of paint raw materials on cellulolytic enzymes present in water borne paint. *Int. Biodeterior. Biodeg.*, **32**, 233–42.

Tothill, I. E., Best, D. J., and Seal, K. J. (1988). The isolation of *Graphium putredinis* from a spoilt emulsion paint and the characterisation of its cellulase complex. *Int. Biodeterior.*, **24**, 359–65.

Cosmetics and pharmaceuticals

Kabara, J. J. (1984). *Cosmetic and Drug Preservation.* Marcel Dekker, New York.

Orth, D. S. (1993). *Handbook of Cosmetic Microbiology.* Marcel Dekker, New York.

4

Built Environment, Structures, Systems, and Transportation

Buildings

INTRODUCTION

The problems of biodeterioration in buildings are many and various. Some relate to the materials used and some to the environment created by the building itself, its associated services, and the uses to which the building is put. Many of the materials used in buildings are already described in other parts of this book. It is, however, impossible to consider buildings without mention of individual materials, and so there is bound to be some overlap. The main emphasis of this section, however, is on the problems of buildings as structures and not as assemblages of materials.

Traditionally, buildings to house humans and their materials have evolved in style and construction with regard to the prevailing climate and within the limits set by available materials and finance. Perhaps today, style has become more important, and it can be argued at the expense of durability and resistance to biodeterioration. It was once said to the authors that there are no biological problems with buildings, only problems of construction. This would be perhaps nearer the truth if all buildings were near perfect in design, construction, maintenance, and ventilation, but we all know, alas, that this is not true.

So what are the main biological problems? For convenience these can be listed and considered under a series of general headings:

1. Fungal and bacterial growth
   a. affecting the structural strength
   b. affecting decorations, paintwork, furniture – 'fouling fungi'
   c. causing health problems.

2. Insect pests
   a. affecting structural strength
   b. infesting non-food contents
   c. infesting stored foods
   d. causing health hazards and aesthetic problems (see Chapter 2).
3. Rodents and birds
   a. causing mechanical damage of fabric
   b. presenting health risks.
4. Lichens, phototrophic microorganisms and higher plants
   a. causing structural problems
   b. causing aesthetic problems.

This classification is purely pragmatic and is not fully comprehensive. Bats, foxes, slugs, and snails have all been cited as having caused problems in buildings, but a comprehensive catalogue of rare or minor examples is outside the scope of this book.

FUNGAL GROWTH AFFECTING STRUCTURES

True dry rot

Most fungi of significance in damaging the structure of buildings (as well as contents when left undisturbed) are the wood-rotting basidiomycetes. In the UK the most serious damage is caused by the true dry-rot fungus *Serpula lacrymans* (Figure 4.1) (known in older texts as *Merulius lacrymans*). The common name of this fungus is confusing, as the wood needs to be damp (as in all cases of decay) before fungal growth and attack can begin. With this species, the wood becomes susceptible at a lower water content (about 25%) than with most other types of decay (compare with 50% for 'wet' rots), and, by producing thick conducting strands of hyphae (rhizomorphs) behind its growing front, the fungus can translocate water (and nutrients) from damp to drier wood. Thus the fungus moves on by transferring water, and in its wake, the decayed wood where hyphae have died back is dry to the touch and scored by characteristic cubical cracking. The vigorous growths of this fungus are able to travel over brickwork and masonry and can penetrate to some extent, enabling them to 'jump' non-wood barriers such as plaster and mortar.

True dry rot thrives best in conditions of static dampness, typical of badly ventilated closed-up properties and cavities. The wet-rot fungi

Figure 4.1. Dry rot: developing fruiting bodies of *Serpula lacrymans* on wall, painted-wood skirting board, and wood-block floor. Photo: Dr Christine Gaylarde.

subsequently considered are much more able to tolerate fluctuating conditions of wood moisture and air humidity. True dry rot is usually difficult and costly to eradicate and involves removal of timbers well into apparently sound wood (at least 2 m), the use of biocides on new timber and brickwork, stripping of plaster, and replacement with plaster containing a fungicidal additive.

Wet rot

This is caused by several different basidiomycetes including *Coniophora puteana* and species of *Fibroporia*. These fungi need high humidity for growth and a substrate moisture content of over 50%. Local pockets of dampness, such as those which may result from the use of impervious floor coverings, faulty guttering, or a leaking roof, may encourage growth to begin, the fungus first appearing as a felt of yellow mycelium. In conditions of continuing high humidity, strands of mycelium may occur which then darken with age, but never attain the thickness of the rhizomorphs of true dry rot. The fruiting body is dull green-brown, with a warty surface. (Dry rot tends to be more yellow and white, ultimately tinged with the colour of the rust-red spores, but fruiting body colour and morphology is very variable and is not totally characteristic, especially in the early stages). Wet rot causes considerable shrinkage of wood, and cracking similar to that of dry rot may occur (this causes some identification problems when

Figure 4.2. Mould growth on a painted church wall in Wiltshire, UK, caused by a leaking drainpipe which is wetting the wall from outside. Photo: Dr D. Allsopp.

only a few wood fragments are available as samples). Treatment is much simpler than that for dry rot, as growth is checked once the dampness is eradicated.

It should now be clear that structural damage by fungi to timber cannot occur in the absence of damp. Dry, well-ventilated buildings do not encourage fungal growth. The difficulty is in spotting sources of damp before any attack occurs.

Fouling fungi

It is common to find mould fungi growing on surfaces within buildings, causing disfigurement by their presence (Figure 4.2) and also inducing changes in decorative finishes by the release of pigments.

The growths are generally superficial and do not cause structural damage, but filamentous fungi have been detected growing within stone (endolithic growth) and a number of fungi produce and excrete enzymes which can degrade the components of wood (see Chapter 2). Fungal growths on walls and ceilings may be the result of condensation forming on colder areas. The level of moisture need not be particularly great; human beings give off moisture, as does cooking. Often, pockets of still air hold slightly more moisture, such as those found behind pictures hung

Figure 4.3. Fungal growth behind a painting hung on a church wall in Salvador, Brazil. Photo: Dr Fatima Bento.

on, or furniture placed close to, walls, producing an atmosphere conducive to mould growth. Figure 4.3 shows such growth behind a painting hung on the wall of a church in Salvador, Brazil. Where growth occurs, better wall insulation, drier air and more ventilation, use of fungicidal washes, fungicidal paint and wallpaper adhesives can all be employed as control measures.

Other mould growth around the house, such as that on books, furniture, clothes, and floor coverings, is a sure sign of generally damp conditions, and immediate steps should be taken to reduce humidity, as remedial treatments for such growth can be difficult, and in advanced cases, practically impossible.

Filamentous fungi are also responsible for discolouration and degradation of external walls. These are most frequently the dark-pigmented mitosporic fungi, and they are generally found in combination with phototrophs (see the next section in this chapter). The substrate (material) of the building is important in determining the rate and extent of colonization. The observer will often notice blocks of grey discolouration on the outsides of painted or rendered buildings. This is fungal (and cyanobacterial) growth over the more porous materials used in the construction of the building, which retain moisture for a longer period (known as time of wetness). The same phenomenon may be noticed on older well-carbonated mortar between bricks and, perhaps more commonly recognized by the general public as mould, on the grouting between bathroom tiles, although the latter may sometimes, in fact, be actinomycetes rather than fungi.

Mature biofilms on external walls contain not only phototrophs and filamentous fungi, but also bacteria (including actinomycetes), slime moulds, protozoa, nematodes, and small animals, such as mites. These latter four organisms can feed on the microorganisms present, either destroying them by digestion, or, when the ingested cells are resistant to digestive enzymes, spreading them over the surface. Figure 4.4 shows an amoebal cell containing undigested algae seen in a sample from a painted wall in Brazil.

Garden furniture is very susceptible to mould and phototroph growth. The fungus *Tripospermum* is found in the "sooty mould" deposits found on plastic furniture left under trees and has also been identified recently in Brazil on external walls painted with an acrylic paint. Storing furniture indoors during the winter does not prevent fungal growth, especially when the environment is humid, but thorough cleaning, twice yearly before and after storage, will extend the useful life of the furniture.

Cold stores present their own problems concerning mould growth. Cold stores vary from 2 or 3 m^2 to large warehouses. The temperature may be slightly above or below freezing depending on the types of goods stored (usually foodstuffs). While in use, the floor and walls soon become contaminated with small amounts of nutrients derived from the stored goods (e.g., fat in meat cold stores). Because of the low temperature, gross mould growth does not develop, despite numbers of spores in the atmosphere of the room. The problems occur when the building is out of use for a time or if the temperature is allowed to increase either deliberately or by accident. Cold stores, by design, do not have heaters, and can quickly become damp enough to allow fungi to exploit the accumulated nutrients. At best, this can present a cleaning problem and at worst, constitute a potential health hazard and spread to any goods still stored.

Cold stores in use are not immune from problems. If moist air from outside enters in any quantity, condensation problems can occur, leading to mould growth, which, although slow because of low temperatures,

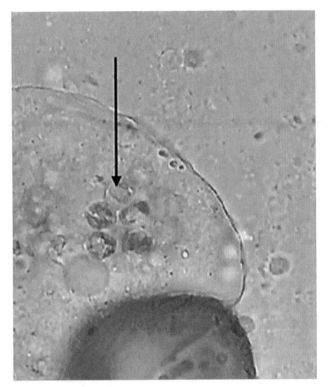

Figure 4.4. Amoebal cell containing undigested cells of the green alga *Chlorella* (arrow). Photo: Dr Christine Gaylarde.

can present a long-term problem. *Cladosporium* has been found growing in cold stores at $-6\,°C$, and other genera of fungi, especially yeasts, have been found growing at temperatures as low as $-23\,°C$. Many fungi can survive temperatures much lower than these, albeit in a dormant state. The inside walls (especially of small units) are often insulated to minimize condensation, but once growth has started, such structures can aggravate the problem by making cleaning difficult, especially if moulds have developed in cavities between the store wall and the insulation panels. The types of biocide and fumigant which can be used in such stores is limited because of possible product contamination, and the most sensible approach is one of thorough cleaning (with suitable detergents and perhaps high-pressure steam jets) followed by drying and checking the cooling equipment. Similar problems have occurred on ships that have been 'mothballed' and even in

Figure 4.5. A grey discoloured wall in Florianopolis, Brazil. Where the structure is protected from rain by overhanging capping stones, microbial growth is reduced. Photo: Dr Christine Gaylarde.

cars sealed for export on long sea voyages, in which ventilation has been minimal or non-existent and changes in the external temperature have encouraged condensation to form inside. The still air allows undisturbed development of fungal growth, which covers walls, carpets, and furniture in an even layer of green mould. This was a particular problem when plasticized PVC was first used as seat coverings in cars.

ALGAL AND CYANOBACTERIAL GROWTH AFFECTING STRUCTURES

The first effect of the growth of photosynthetic microorganisms is disfigurement (discolouration) of the surface, often mistaken by the layperson for simple 'dirt' (Figure 4.5). Coated surfaces, wood, plastic, concrete, and natural stonework can all be affected. Growth is rarely uniform; frequently there are streaks of growth, grey, black, brown, green, or orange, indicating areas of dampness or water run-off. The lower portions of external walls

Figure 4.6. A restaurant in Aguas Calientes, Peru. There is intense green discolouration at the base of the wall. Photo: Dr Christine Gaylarde.

are often particularly affected, as they remain wetter than upper regions (Figure 4.6). Leaking pipes and areas beneath the condensate drain from air conditioners are also good examples that are readily seen (Figure 4.7). More homogeneous biofilms are particularly common in the humid tropics, where they may be especially apparent in more shaded areas, as very high levels of UV light are inhibitory.

Once established, growths may form a slippery film, for example, on steps and paths, which can constitute a danger to pedestrians. More direct damage can be caused to the structure itself by the acids and other metabolites produced by the organisms, which, together with the ability of some species to tunnel into surfaces, lead to degradation, increased porosity, and decreased durability of the structure. Figure 4.8 shows the blue-green growth revealed by spalling of the soapstone surface on a church in Minas Gerais, Brazil. This endolithic growth (most frequently cyanobacteria or fungi) is particularly damaging to the structure of a building.

Algae differ from cyanobacteria in their cellular structure, physiology, mode of reproduction, and resistance to environmental factors. It is important to distinguish between the two types of phototrophic microorganisms, as different biocides may be needed for their control. In spite of this, both types of phototrophs were regarded in the past as

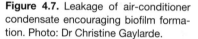

Figure 4.7. Leakage of air-conditioner condensate encouraging biofilm formation. Photo: Dr Christine Gaylarde.

algae, and both occupy more or less the same niches on built structures. They are found mainly on external surfaces, but can also be important causes of biodeterioration on internal structures, as long as a certain amount of light impinges on the area. Cyanobacteria, in particular, can grow at low light levels and even, occasionally, in the absence of light, metabolizing preformed organic matter. On important historic buildings of the Maya culture, in the Yucatan peninsula, Mexico, much thicker biofilms were found on internal than external walls. The biofilms contained, as their major components, cyanobacteria and, to a somewhat lesser extent, fungi. These microorganisms grew more luxuriantly in internal sites protected from high UV levels and with a higher humidity. Droppings from birds and bats also added to organic carbon levels on these surfaces.

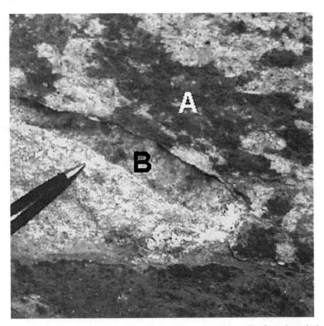

Figure 4.8. Spalled soapstone on the external wall of a church in São João del Rei, Brazil. A, red patina on surface; B, blue-green biofilm exposed below recently spalled stone. Photo: Dr Christine Gaylarde.

As phototrophic microorganisms require only sunlight, water, and minerals for their growth, they are normally thought to be the primary colonizers of buildings, paving the way for the heterotrophic bacteria, fungi, and higher organisms by providing a layer of organic food. However, a recent study in Brazil suggests that this is not the case. On newly painted external walls, fungi were detected on the surfaces before phototrophs began to colonize. These organisms could be using dirt deposited from the atmosphere, or organic components of the paint itself, as carbon sources.

CONTROL OF MICROBIAL GROWTH

Control of fouling by fungi and phototrophic microorganisms, where not possible by alteration of environmental conditions (such as reduction of humidity and/or incident light), is generally attempted by use of biocides (see Chapter 6). Algicides and fungicides can be incorporated into liquid coatings to prevent colonization of the dry film for the effective life of the coating (generally 5–10 years).

Where painting or varnishing is not possible (for example, with some historic and cultural buildings), simple cleaning, sometimes with the use of biocides, is the only option. High-pressure water jetting is normally used and will generally remove most of the biofilm, but care must be taken to avoid removing the surface of the underlying material! A mild biocide, such as hypochlorite, is often used in addition, and this reduces the residual microbial populations and may help to prevent regrowth while the surface dries. Care must be taken, however, not to increase the salt levels in the material, as this in itself can be damaging. Thorough cleaning is also recommended before repainting, as painting over a contaminated substrate is simply throwing money away. A treatment of 15 min with hypochlorite, followed by the water jet (11 MN/m^2), reduced the fungal, algal, and cyanobacterial populations on two painted buildings in Sao Paulo, Brazil, by over 98%.

HEALTH PROBLEMS

Potential and actual human health problems (other than food contamination, which is dealt with elsewhere) can occur because of microorganisms in buildings. Fungal spores, dead or alive, and the microinvertebrates such as mites which browse on fungal material, especially when present in large numbers, can trigger allergies, such as allergic rhinitis, or induce asthma. Such biological factors are involved in sick building syndrome, a complex situation in which occupants experience a variety of symptoms and become generally unwell, recovering only when they cease to frequent the building. Other factors include dust in the air, volatiles from decorations and furnishings, and poor ventilation. In the United States in recent years there have been cases of serious lung conditions (especially in infants) and other conditions, suspected to be caused by the presence of large growths of the fungus *Stachybotrys chartarum (S. atra)*. This fungus grows particularly well on cellulose-based building boards commonly used in some areas. When such materials become damp, usually because of condensation and poor insulation and ventilation, growth occurs and spores are circulated in integral air-conditioning and ducted heating systems. Cases have led to expensive litigation, and draconian measures have been taken, including evacuation and even demolition of buildings. New remedial protocols have been drawn up, such as the *Guidelines on Assessment and Remediation of Fungi in Indoor Environments* by the New York Department of Health, which set new standards for treatment with corresponding increases in cost. These guidelines are beginning to be adopted by remedial contractors in Europe, but problems of assessment of the actual risks remain to be fully resolved. All fungi are considered equally hazardous, and risk is primarily assessed on area of growth, irrespective of density of growth or amount of sporulation. Numbers of spores or propagules in the atmosphere are not routinely counted or used as a hazard indicator except in cases in which clinical symptoms are present or ventilation systems are involved. Refinement of the criteria would seem to be required.

Liquid systems in buildings can also give rise to problems. Water systems in industrial cooling plants, air conditioners, dehumidifiers, showers, and recirculating spa baths (jacuzzis) can become sources of infection if not regularly and effectively cleaned and disinfected. The bacterium *Legionella pneumophila* growing in such systems has led to a number of well-publicized fatalities worldwide caused by 'Legionnaire's Disease', including those in the UK in the summer of 2002 and the autumn of 2003. (The first cases were at a Philadelphia convention of the American ex-servicemen's association, the American Legion, in 1976.).

INSECTS AND STRUCTURES

The insects which inhabit buildings can cause a variety of problems. Some are casual visitors from outside, causing little more than a temporary nuisance, others are more intimately concerned with human activities, living permanently within buildings, attracted by the availability of food, harbourage (shelter), and warmth; A third broad category of insects are those that not only live within the structure, but also often utilize it as a nutrient source, eventually causing physical damage.

Casual visitors include 'garden' insects and other invertebrates such as bees, wasps, earwigs, and even slugs and snails. Their entry is usually seasonal and may be influenced by freak weather conditions. One factory visited by the authors had reported spotting of ceilings that was due to mould growth. The spots proved to be insect remains in dried-out spots of condensation. Both the condensation problem and the influx of insects from nearby agricultural land were seasonal.

Commensal insects, for example cockroaches, give rise to greater objections. Their mere physical presence is unwelcome, and their consumption and contamination of foodstuffs and their potential to spread disease makes control essential. The building itself is threatened by other groups of organisms, those which burrow into the structure, to seek harbourage or spaces for egg laying and larval development, and, in the case of wood-boring organisms, food material for larvae; the main types of organisms in this last category are considered in the next subsection. (See also Chapter 2, the section wood in the marine environment.)

Wood-boring insects

Woodworm (*Anobium punctatum*)

Of all premises in the UK suspected of having wood-boring insect infesta-tions, over 70% are infested by *Anobium punctatum*, the common furniture beetle or woodworm (it is the larva which bores into the wood). There has been a gradual decline in active infestations of *A. punctatum* in houses over the past 30 years. This is because the eggs and young larvae require a higher moisture content in the wood to survive. Widespread use of central heating has reduced relative humidities to low levels. In the natural state, this insect colonizes dead trees, but in premises it is found in structural timbers and furniture, and not necessarily in old wood. Plywood can also be affected, and the old-fashioned forms which used animal glue in their make-up were particularly nourishing, enabling the insect to complete its life cycle in approximately 1 year rather than 3.

For most of its life cycle, the insect is not visible and infestations are usually detected by small neat flight holes in wood. The presence of holes does not necessarily indicate an active infestation. Frass produced from holes may give a clue, but investigators can keep a check only by mark-ing flight holes and waiting for any increase as pupae hatch and adults emerge.

Figures 4.9 and 4.10 show typical damage due to woodworm.

Death-watch beetle (*Xestobium rufovillosum*)

Probably accounting for only about 5% of all insect infestations in the UK, and then only in particular types of wood not commonly used in modern buildings, the death-watch beetle has a place in the public mind disproportionate to its frequency. It is a pest of hardwoods such as oak beams and is hardly ever found in sound, dry timbers. These factors tend to make it a pest of old and historic buildings, and this gives it its claim to fame. It takes its common name from the fact that the adults tap on wood to attract mates. The life cycle is usually longer than that of *Anobium*, taking 4–6 years. There appears to be a strong link between fungal decay and speed of development of the larvae (a factor also relevant to other wood-boring species), although the relationship may involve several factors, including physical and nutritional ones.

Figure 4.9. Woodworm (*A. punctatum*) damage to wood. Photo: Dr David Pinniger.

Figure 4.10. Active woodworm infestation in wood. Note cones of fresh bore dust beneath flight holes. Photo: Dr David Pinniger.

Powder post beetles (*Lyctus* spp.)
In the UK, powder post beetles are represented by *Lyctus brunneus* and three other similar species. *Lyctus* cannot use the cellulose of wood; it feeds on starch, and is therefore found only in the sapwood of certain hardwoods. It is a pest of relatively new wood (up to 14 years old). The wood is destroyed in layers, similar to termite damage, and a thin veneer of sound wood is left on the surface, protecting the insects and disguising the true extent of the damage. The boredust is very fine, and from this the beetles get their name. At higher temperatures, the life cycle is completed in less than 1 year, resulting in rapid onset of infestation. There seems to have been an increase in the incidence of *Lyctus* in Southern England in recent years.

House longhorn beetle (*Hylotrupes bajulus*)
The house longhorn beetle is a major pest in Scandinavia. In the UK it was for many years restricted to certain areas in the south of England; however, it is now more widespread. In areas where attacks are more common, building regulations require chemical treatment of timbers used in new constructions. The insect is usually found in softwood roofing timbers, where the large larvae (up to 24 mm long) eat into the sapwood, leaving a surface veneer, which can make detection of an infestation difficult. The flight holes of the adult beetle are oval. The life cycle is long, between 5 and 10 years.

Weevils
Wood-boring weevils, such as *Pentarthrum huttoni* and *Euophyrum confine*, are small dark beetles, usually found in damp timber, and are associated with fungal decay. They consume the softer portions of wood (springwood) first and both the adult beetle and the larvae bore. In the days when wooden beer barrels were common, *Pentarthrum* may have been introduced into the damp timbers of cellars from the wood of the barrels. The life cycle of these weevils is short, usually less than 1 year.

Control of wood-boring beetles
The control of wood-boring beetles in structural timber is a contentious subject. Many timbers in houses contain residues of past treatments with lindane (gamma HCH) and pyrethroids. A large number of treatments with residual pesticides are still carried out on timber with beetle flight holes. This is often to meet the requirements of a building survey. In many cases the damage is old and the infestation is no longer active and may have been

dead for years. It is essential to try to determine if infestation is currently active by an examination for fresh beetles and emergence holes and frass falling out of flight holes. Most wood-boring beetle emergence takes place in spring.

If there is activity, then a residual insecticide, such as permethrin, can be applied to the surface of the timbers. This will kill emerging adults and larvae near the surface but is unlikely to kill larvae deep in wood. Control may therefore take some years. It is pointless trying to control woodworm and death-watch beetle with pesticides unless the environment is improved to make the wood less suitable to prevent them continuing the infestation. Control of insect infestation in materials and buildings generally, by use of repellents (probably in conjunction with other measures), is a topic which may develop in the future.

Termites

Although termites are not a problem in the UK and other temperate regions, the significance of their effects on materials on a world scale is so great as to demand their inclusion in any introductory account of biodeterioration. They are the main wood-destroying insects in the tropics and sub-tropics, and there is a thriving industry devoted to their control, especially in developed regions.

Termites are inconspicuous creatures, requiring high temperatures and humidities, which remain hidden from view for much of their lives. They are one of the oldest insect groups. All termites are social insects and have a well-developed and complex colonial structure. The different castes and stages of development are given in the following subsections.

WINGED ADULT The active flying adult stage of the termite is extremely brief. Male and female winged adults swarm from the parent nest, and, after mating, the pairs shed their wings and become reproductives. These are known as kings and queens and form the nucleus on which a new colony is built.

WORKERS These are young stages which maintain a nymphal form but with strongly chitinized mandibles similar to those of adults. The development of nymphs which become workers is arrested after the second moult.

SOLDIERS Soldiers have a large head to house the musculature of the formidable mandibles; the rest of the body remains very similar to the nymphal form. They are wingless and sterile.

DEVELOPMENT The type of development shown by termites is hemimeta-bolous, or undergoing incomplete metamorphosis. When the young hatch, they already have mouthparts and legs, although the body is soft and there are no wings. There is no pupal stage as seen in the Diptera (true flies), Lepidoptera (butterflies and moths), and other homometabolous insects. There are thus three stages in the reproductive cycle: the egg, the nymph, and the winged adult or imago.

The small eggs are cylindrical, produced in large quantities by the queen. The insect moults several times during its nymphal phase, during which the wings develop and size increases. At the final moult, reproduc-tive organs develop. Unlike many other insects, the adults may live for several years, even though fully developed wings are present for only a short time to allow dispersal to take place.

Being social insects, termites live and develop in an environment of their own making; the colony influences the development of each nymph in accordance with immediate needs. This termite 'social engineering' puts the needs of the colony before those of individuals, and thus nymphs, rather than progressing smoothly through to winged adult and potential reproductive stages, may be diverted into other developmental paths, to become soldiers or sterile workers. Even when development has taken place, it can to some extent be reversed, in that if winged adults mature at a time unsuitable for swarming, they may literally have their wings clipped by other termites and be retained in the nest to act as extra workers.

FEEDING Some species of termite utilize organic matter in soil or culture fungi in their nests. Those of main interest in biodeterioration are the ter-mites which utilize cellulose, usually as wood. The alimentary canal is a relatively simple tubular structure, incorporating several features to aid the use of wood as a foodstuff. There is a chamber, the muscular proventricu-lus, lined with hard chitinous plates, where food material is ground up, and there are symbiotic protozoa present in the hind gut which produce cellulase and enable the wood to be digested. These protozoa are complex flagellates, the most common being members of the genus *Trichonympha*.

WORLD DISTRIBUTION There are only two species of termite native to South-ern Europe, *Reticulitermes lucifugus* and *Kalotermes flavicollis*, (although *R. flacipes* has been found indoors in Hamburg, in France near Paris, and one established infestation has been found in South-West England). The majority of termites are found in the warmer and more humid parts of the world, including North Africa and the Middle East, Africa south of the Sahara, tropical America and Asia, Malaysia and the East Indies and Australia. Termites are spread by human commercial activities, in timber and wooden articles, although the number of species able to survive such movements depends on the dampness of the wood, with most termites preferring moist conditions.

DAMAGE CAUSED BY TERMITES Termites can damage a wide range of agricultural crops, eating roots and causing soil erosion. Growing trees can also be badly affected, either as seedlings in plantations or as mature trees in plantations or natural forests. These effects, however, are outside the scope of this book; only timber is considered here. Unlike the termites which build large soil mounds as nests, many wood-destroying termites construct nests inside timber (Figure 4.11) removing much of the internal wood, leaving only a thin veneer on the surface to insulate the nest from drying and from predators. These are the drywood termites; other wood-destroyers live in underground nests and travel to sources of wood, carrying back fragments to consume in the nest. Such termites usually enter the wood of buildings from soil tunnels below the structure and are therefore difficult to detect. Where they approach on the surface they may construct tubes or runways of wood and soil fragments to span gaps or cross stone and concrete barriers. In these cases, the termites are easily seen. The cost of such damage is high; in the United States and Australia, the cost is in the region of tens of millions of dollars each year, and in the tropics the use of wood as a building material may necessitate annual maintenance costs of 10% of the property value each year.

Control of termites and protection of buildings
1. *Mechanical* methods can be used, with varying degrees of success. To be of maximum benefit, frequent inspection is needed and runways should be removed as they appear. Buildings can be erected on a concrete raft which extends well clear of the edge of the structure, alternatively, or in addition, the structure can be built on hardwood or masonry pillars capped with metal non-corroding skirts, 'ant-guards', which deter runway formation. Another variation on this theme is to incorporate a concrete slab, well clear of the soil, between conventional foundations and the floor of the building.
2. *Chemical* barriers generally consist of soil poisoning on a massive scale. DDT, chlordane, dieldrin and sodium arsenite, pentachlorophenol, and

Figure 4.11. Insect attack on wood. Remnants of wooden cupboard door after attack by drywood termites (*Cryptotermes brevis*). Photo: Dr Neiva Barros, University of Caxias do Sul, Brazil.

copper naphthenate have all been used. Effectiveness depends on soil type, exposure to leaching, and the persistence of the chemicals. There are obvious potential problems with this approach in terms of environmental pollution. Toxic baits containing fipronil or hexaflumaron have been developed for termite control. These require expert knowledge and careful placement for their successful use.

3. *Fumigants and wood preservatives* may also be used, as drywood termites may not arrive via the soil or runways, but as winged adults, nesting subsequently in the wood well above the ground. With fumigation, the whole building is treated after covering and sealing, a process which takes several days, involving the removal of people, pets and plants. (See also the section on historic buildings in this chapter.)

RODENT AND BIRD DAMAGE

Rodents in buildings

Pest organisms cannot be considered in isolation from their habitat. Often, the habitat and the food source are essentially the same; hence wood-boring insects are considered mainly under the sections on wood in Chapter 2. Our interest in rodents, however, owing to their effects on humans and to their omnivorous eating habits, makes this chapter on buildings the most suitable place to consider them as deteriogens; indeed, the name house mouse (*Mus musculus/domesticus*) makes the point well.

Animals are given names by humans, however, and as will be seen, the activities of rodents are not confined to dwelling houses; the significance of rodents as deteriogens is in their activities as commensal organisms (i.e., living in association with humans).

The rodents are a large group of mammals with some 2000 representative species worldwide. The word 'rat' may properly apply to any of about 500 species of animal and the word 'mouse' to at least 130 species.

Although there are many rodents, relatively few worldwide live in very close association with man to the extent that thay may be classed as commensal rodents. Those commensals that have a worldwide distribution are the common, brown, or Norway rat (*Rattus norvegicus*); the black, ship, or roof rat (*R. rattus*); and the house mouse (*M. musculus/domesticus*) (see Table 4.1). Those with more limited geographical range include the multimammate rat (*Mastomys natalensis*) in Africa, the lesser bandicoot rat (*Bandicota bengalensis*) in the Asian subcontinent and the Polynesian or Burmese house rat (*R. exulans*) in the Pacific and Asia.

Rodents are most easily distinguished from other mammals by the characteristic arrangement and form of their teeth. They have only a single pair of incisors in both the upper and the lower jaws and no canines. The wide gap (diastema) between the paired incisors and the molars (or cheek-teeth) gives the rodent skull an unmistakable appearance.

The incisors form the clue to the tremendous success of rodents within the animal kingdom. Rodent incisors have three basic characteristics that together distinguish them from the teeth of most other animals: They are strongly curved, they grow continuously throughout the animal's life, and they carry a layer of enamel on the outside surface only. The fact that rodent incisors grow continuously means that they can also be worn away continuously. The gnawing action results in the softer dentine's being worn more rapidly than the hard enamel, giving a chisel-like outer edge to the teeth, which can penetrate soft metals such as lead and aluminium.

Diseases carried by rodents

Of the many diseases carried by rats and mice, plague at once comes to mind. This is spread from rats to humans by fleas. Plague is still endemic in many areas, and has been found in the south-western United States, South America, Democratic Republic of Congo, South Africa, Kenya, Madagascar, India, Java, North China, Iranian Kurdestan, and parts of the former USSR.

Table 4.1. Characteristics of rats and mice

Character	Norway rat (*R. norvegicus*)	Ship rat (*R. rattus*)	House mouse (*M. domesticus*)
Average adult weight	About 330 g (a specimen of 725 g is known)	Less than 250 g (a specimen of 360 g is known)	Usually less than 25 g (average 15–16 g)
Tail length	Shorter than head and body	Longer than head and body, except in some foreign forms	Usually longer than head and body
Ears	Thick, opaque, short with fine hairs	Thin, translucent, large, hairless	Large, some hairs
Snout	Blunt	Pointed	Pointed
Colour	Brownish grey, but may be black: grey belly	Grey, black, brown, or tawny; may have white belly	Brownish grey; lighter shades occur
Droppings	In groups, but sometimes scattered; spindle shaped or ellipsoidal	Scattered, sausage, or banana shaped	Scattered, thin, spindle shaped
Habits	Burrowing, can climb; swims; gnaws; lives outdoors, sometimes indoors, and in sewers	Non-burrowing, agile climber; gnaws; lives indoors in Britain; rare in sewers	Climbs; sometimes burrows; gnaws; lives indoors and outdoors in warm climates; almost unknown in sewers

However other rat-borne diseases may be of more practical concern, for example, leptospiral jaundice (Weil's disease). The causal organism, *Leptospira icterohaemorrhagiae*, is excreted in rats' urine. Human infection can occur from contact with wet, rat-infested surfaces, from accidental immersion, or from bathing.

Trichinosis, a disease associated with encystment in human muscles of a small parasitic nematode worm, is carried by rats and pigs acting as intermediate and reservoir hosts. Bacteria of the *Salmonella* group, which can cause food poisoning, can be carried in the excreta of rats and mice. Food poisoning is perhaps the most common human disease carried by rodents.

There are two illnesses called 'rat-bite' fever. Both involve swelling of lymph glands and muscular pains, and relapses may occur long after apparent recovery. Mice also can carry these infections. Murine typhus and scrub typhus are both rickettsial diseases carried by rats, the causal organisms being transmitted to humans by rodent fleas or by mites. A rather uncommon but serious viral infection of human beings (especially children) known as lymphocytic choriomeningitis is carried by house mice. Instances are known of house-mouse carcasses causing tularaemia (a plague-like infection) in persons handling them. Mice also spread, through their droppings, the tapeworms *Hymenolepis nana* and *H. diminuta.* Favus, a skin disease of fungal origin, may be contracted from mice or indirectly through cats. The Hantaan virus is carried by a range of rodent species and is transmitted to humans through environmental contamination by rodent faeces, urine, and saliva.

Many additional diseases, including foot-and-mouth disease, may be transmitted by rodents.

Losses caused by rodents

It is common knowledge that rats and mice damage foods, crops, and buildings, including structures such as sewers. The costs of rodent contamination by faeces, urine, hair, and carcasses may also be severe. It has been estimated that 12%–15% of world food production is lost to rodents. Damage to structures by gnawing is widespread. Almost any kind of material may be attacked. The enamel of the outer surface of rats' incisors is very hard (5.5 on the 'scratch scale' of Moh), whereas the dentine behind is somewhat softer (about 3.5) and thus a very efficient chisel-like edge is always present. Lead, aluminium, and concrete may all be damaged by rodents. Rodents frequently damage electrical wiring, causing electrical failures and fires, and also lead and plastic pipework, causing leaks and flooding. Damage to agricultural primary production may also be severe, particularly in less developed regions.

Biology and behaviour of commensal rodents

Habitats
Rats and mice are quick to take advantage of cavities in the walls, roof spaces, and ducts of buildings, for harbourage. In stockpiled foods, they frequently nest in the crevices between sacks and, where possible, in the

Table 4.2. Breeding potentials

Parameter	Norway rat	House mice
Life expectancy	10–12 months	8–10 months
Litters per year	Up to 5	About 6
Gestation period	21–23 days	19–21 days
Litter sizes	2–14 (average 8)	2–13 (average 6)
Eyes open	9–14 days	11 days
Eat solid food	21 days	11 days
Mating age	8–12 weeks	6–10 weeks
Breeding season	All year in warm climates. In temperate climates peak in summer	Throughout the year when living indoors
Oestrus cycle	Every 4–5 days	Every 4 days

sacks themselves. Mice can harbour in crevices and holes in cultivated and uncultivated ground and in hedgerows and undergrowth. They also occur regularly in newly harvested cereal and hay crops that are stored in the open.

The jumping and climbing capability and relatively small size of these species allows them to range freely in most environments and to find safe harbourage.

Breeding and development

Under optimum conditions of climate, surplus food, free water, and shelter, rats and mice may breed throughout the year.

The breeding potentials of some commensal rodents are summarized in Table 4.2. Once established in rich environments, rodent populations can increase rapidly.

Senses

Rodents have very highly developed senses of taste, touch, hearing, and smell. Although their eyesight is less well developed, they are very good at detecting movement (predators) within their range of sight.

Food

All three species in Table 4.1 are omnivorous and are capable of feeding on a wide range of foods. All species sometimes supplement their diet by eating such items as worms and crabs, and occasionally take unusual substances such as soap, glue, plaster, and putty. The black and Polynesian rats may prefer fresher food.

House mice can survive without free drinking water and can obtain the water they need from the food they eat. The Norway rat is most prone to suffer from water shortage. Under extreme conditions of water deprivation the fertility of all commensal rodent species declines. Indoors, rats and mice may obtain water from uncovered fire buckets, dripping taps, or leaking roofs, sources which can all be made unavailable to them.

Movements

The movements of rats and mice are largely determined by climatic conditions and the availability of food, water, and harbourage. Mass migrations occur infrequently and are usually associated with sudden and drastic changes in environmental conditions such as crop failures or floods. Seasonal movements of Norway rats and mice occur regularly, however, in countries that have marked seasonal climates. Immigration to new areas is also encouraged by territoriality and increases in population density beyond the carrying capacity of the infested environment.

Although rats sometimes travel considerable daily distances between harbourage and food, extensive wanderings are not the rule. In situations in which food and harbourage are adequate, rats and mice tend to have a restricted range and to follow regular routes. Their ranges tend to be smallest when they are living at high densities in such localities as food stores where food and cover are coincident.

Norway rats regard strange objects placed in their surroundings with suspicion and investigate them cautiously. The term 'new object reaction' or 'neophobia' has been applied to this behaviour.

The signs and traces left behind by rats and mice that are of most use in determining their distribution and relative abundance are droppings, runways, smears, holes, and the damage caused to foodstuffs, packing materials, books and the fabric of buildings.

Individuals tend to follow much the same route when travelling inside or entering a building, so that dark-coloured smears gradually develop along well-travelled runways, around holes, and along beams and girders, where the grease and dirt on their fur has rubbed off. In places where mice have been present for a considerable time, it is also sometimes possible to find small mounds called 'urinating pillars' which consist of a mixture of droppings, dirt, grease, and urine.

Control of rats and mice

Rodents have probably been controlled by human beings since they first evolved. The history of rodent control shows that, in very early times, curses and adjurations were thought effective and the practice of rhyming rats to death is mentioned by Shakespeare. In the reign of Queen Elizabeth I a bounty was offered of one penny *'for the heads of everie three rattes or twelve myse'*.

Poisoning

Until the early 1950s only 'acute' (single-dose) poisons were used for rodent control. At one time the most common way of using these was by 'direct' poisoning, that is, by simply putting down baits containing the poison without any preliminary attempt to condition the rat or mouse colony to feed. This method does not consistently give a very high percentage kill, for if rodents feed so slowly on a poisoned bait that they experience unpleasant symptoms and cease feeding before having taken a lethal dose, they may be difficult to poison again. Often they will not accept the same bait or poison again for several months. This phenomenon has been called 'shyness'. To overcome it, the system of poisoning known as 'prebaiting' is used.

Some of the commonly used acute rodenticides available around the world include alpha-chloralose, fluoroacetamide, scilliroside (red squill), sodium monofluoroacetate, thallium sulphate, and zinc phosphide. Sub-acute rodenticides include bromethalin and calciferol.

In 1950 a great advance occurred in rodent control. This was the introduction of the 'anticoagulants'. The success of the anticoagulants as rodenticides depends on the fact that, when they are eaten by rats and mice at low concentrations in bait, symptoms of illness are slow to appear and the animals do not therefore associate them with their food. Thus low-level feeding continues daily until a lethal dose has been absorbed. At the strength at which anticoagulants are used, the bait may need to be eaten for several days before this happens (and death may occur later still), hence the use of the term 'chronic' poisoning to describe the control technique.

Chronic poisons are safer than acute poisons where other animals are concerned, as a single feed of bait is usually (but certainly not always) insufficient to kill.

Anticoagulants interfere with the production of a substance known as prothrombin, which is necessary for blood to clot quickly when blood vessels are broken. Animals poisoned with anticoagulants may therefore die of haemorrhages resulting from the minor damage to blood capillaries that occurs in the rough and tumble of everyday activity.

The prothrombin content of the blood after anticoagulant poisoning may be restored by dosing with vitamin K_1 which is therefore an antidote.

There are now a wide range of anticoagulant rodenticides available worldwide. First-generation anticoagulant rodenticides include warfarin, pindone, chlorophacinone, coumachlor, diphacinone, and coumatetralyl. Second-generation anticoagulants include difenacoum, bromadiolone, and difethialone. Flocoumafen and brodifacoum are employed for pulsed baiting (see subsequent discussion).

The use of chronic rodenticides is based on a technique known as 'saturation baiting'. This means that the infestation is thoroughly surveyed and bait points laid to cover the infestation effectively. Revisits are then made to the bait points frequently enough to ensure that there is always bait available to the rodents. In this way the rodents have regular (daily) access to the bait points even though they may not feed extensively at any one point or on any one day.

The development of the two newest anticoagulants (brodifacoum and flocoumafen), however, has brought with it the opportunity to introduce a new baiting technique; this has been termed 'pulsed baiting'. These two anticoagulants have rodent toxicities that are higher than the other anticoagulants, so high in fact that it is easier for a rat or mouse to ingest a lethal dose at a single feed; death is still delayed, however, and no extreme symptoms are apparent; thus no poison or bait shyness develops.

The higher toxicity of brodifacoum and flocoumafen has resulted in their use being more strictly controlled in some countries.

It is essential that the labelled instructions on the container for the use of the rodenticide formulations are followed.

Most rodenticides are used in edible bait formulations; alternative methods of presentation are, however, available; these include

1. contact dust formulations
2. liquid formulations
3. contact gels
4. contact wicks.

Anticoagulant (physiological) resistance

Resistance among both house mice and rats was not slow to develop and was causing problems with control in restricted geographical areas, particularly Northern Europe and North America, by the end of the 1950s.

Current geographical spread is not easy to determine, although there is more widespread resistance to the first-generation than to the second-generation anticoagulants. Anticoagulant resistance may be recognized by the continued uptake of the baits with no corresponding decrease in activity. In other words, some populations of rodents continue to consume the anticoagulant baits without harm.

Other methods of rodent control

In addition to the widespread use of rodenticides as the main method of reducing rodent populations, it is important to be aware of a range of alternative methods of preventing rodent problems; these include

1. traps
2. ultrasound
3. fumigation
4. hygiene
5. proofing.

TRAPPING Traps may be used for rodent control, although to be effective they are best used intensively when controlling significant infestation. Both live traps and kill traps are available. Rodents may avoid traps, particularly if they have been hurt or frightened by them.

ULTRASOUND These devices utilize the rodents' ability to hear ultrasound as a means of frightening them away. The efficacy of these devices is, however, not proven.

FUMIGATION Fumigation and the use of gasses to control rodents is very hazardous and should be used only in exceptional circumstances, and then only by licensed specialist pest-control contractors.

HYGIENE AND PROOFING In general, the two factors that are of most importance in controlling rat and mouse populations are food and shelter. If one or the other can be eliminated the rodents must go elsewhere, hence the importance of strict attention to hygiene. Whenever possible, food should always be kept in rat-and-mouse-proof containers. Edible refuse and 'empty' food tins should be placed in bins with tightly fitting lids.

Where food has to be stored in bags or other containers vulnerable to rodents, it should be stowed in such a way that it can be inspected at frequent intervals, and the building in which it is housed should be proofed. Piles of rubbish, timber, bricks, and other materials should not be allowed to accumulate – either indoors or outside – if rats and mice are to be denied harbourage. The rodent population in Chicago has been greatly reduced in recent years by the introduction of lidded refuse bins ('wheelie bins') which replaced plastic garbage bags.

Proofing buildings against rodent access is the best means of preventing problems. Rodents are physically very able, with the ability to perform the following feats:

- To climb vertical wires and ropes of 2–3-mm diameter or more
- To pass through small holes. Young Norway rats can pass through holes larger than 1.25-cm diameter and young mice through holes of 5-mm diameter (the diameter of a pencil)
- To climb inside vertical pipes 4–10 cm in diameter
- To climb pipes within 7.5 cm of a wall or other continuous surface
- To jump vertically (black rats, 1 m; Norway rats, 60 cm; house mice, 30 cm)
- To jump horizontally over 1 m from a standing start
- Norway rats can dive through water traps in foul water systems and swim significant distances.

Entry of rats and mice into buildings may thus occur in several ways. One route is from the sewers through a drain. One may prevent this by ensuring that where an interceptor trap is fitted the water seal is effective and that the caps of any rodding arms are in position. Entry by climbing up the inside of ventilation and rainwater pipes may be countered by the installation of wire balloon guards at the top.

Access to rooftops by the use of the outside of vertical pipes close to walls can be circumvented by 20-gauge metal pipe guards fitted tightly to the pipe by an adjustable metal collar and projecting about 22 cm. Cone guards must lie snugly against the wall, whereas square guards are best built into a brick joint and should have the edges turned down about 5 cm for strength.

Other ways in which rats and mice enter buildings are by climbing the face of rough stonework and brickwork, climbing creepers and other decorative foliage, through defective doors, windows, broken airbricks and ventilators, by tunnelling under walls and through foundations, and by walking along telephone and other cables and the branches of trees.

Ventilation grids and airbricks should be proofed externally either with
5-mm mesh, 24-gauge expanded metal, or with galvanized-steel-woven
wire cloth. Doors and windows should be close fitting, and any broken
panes should be replaced. Windows, fanlights, and ventilators that are
permanently open should be meshed over, as described for airbricks.

Birds and buildings

Birds are a natural component of the urban habitat. However, the construction of buildings within these urban habitats favours some birds more than
others. Similarly, the use to which these buildings are put may favour some
species more than others, particularly when food production, processing,
storage, or sale are involved. In these circumstances, the combination of
structural diversity and the availability of spilt, waste, or discarded food is
particularly attractive to a number of species.

Those species particularly favoured include the feral pigeon (*Columba
livia*), the house sparrow (*Passer domesticus*), starlings (*Sturnus vulgaris*),
and some gull and corvid species.

Although attractive to many people, these birds, when concentrated
in large numbers and/or using limited areas particularly intensively, can
cause the following very serious problems:

- Transmission of diseases. A wide range of diseases, including those
  caused by *Salmonella* spp., *Listeria* spp., *Campylobacter* spp., *Chlamy-
  dia psittaci*, *Histoplasma* spp., and others can be carried and transmit-
  ted. The mobility of the bird species means that they are susceptible
  to picking up diseases from a wider range of habitats than less mobile
  terrestrial mammals.
- Initiation of allergic responses in humans
- Noise contamination
- Smell
- Structural damage
- Structural contamination and fouling
- Carriage and transmission of insect pests
- Dangers from surface contamination and bird strikes.

Control

Historically, most emphasis was placed on the culling of birds to reduce
numbers as a means of alleviating the problems. However, it was evident

that, if no actions were taken to remove the factors that attracted the birds to a particular site and if nothing was done to remove the food source that both attracted the birds and enabled them to breed more effectively, the problem returned very rapidly. The mobility of the birds enabled rapid migration into the site.

Thus, although culling methods such as poisoning, trapping, and shooting still have a part to play in bird control, the emphasis is now placed on environmental management to reduce bird problems.

In most areas of the world there are also strict legislative controls on actions that may be taken against birds and the use of culling techniques is often limited.

ENVIRONMENTAL MANAGEMENT Food availability is often the root cause of a bird problem, and much emphasis is now placed on removal of the food source that may be attracting the birds to the site.

This is usually achieved through improved hygiene and cleaning schedules where possible. However, where the public are feeding the birds for personal, religious, or other reasons, this is often not easy to achieve. Thus much emphasis is now placed on the use of exclusion and deterrent and repellent strategies for preventing access by the birds to either their food source or the areas where they may be causing problems.

The following proofing techniques are currently applied:

- Netting to prevent bird access
- Antiroost devices to prevent birds settling on specific preferred areas (gels, sprung-wire systems, spike deterrent systems)
- Visual scaring
- Audio scaring by either random-noise or species-specific distress calls.

Not all the preceding techniques work equally well and are often site specific and species specific in their application. In addition, much knowledge of both the technique and the bird biology and behaviour is required for achieving maximum efficacy.

Bats in buildings

Colonies of bats may select roof spaces in which to roost and can pose problems that are due to an accumulation of droppings. Large quantities of droppings can cause collapse of ceilings and can become sources of smells and breeding grounds for unwelcome insect pests. Bat faeces may carry potential pathogens such as *Histoplasma*, and bats themselves can carry diseases. A fatal case of a rare form of rabies occurred in Scotland in 2002 after a bat conservationist was bitten by an infected bat, the first such case in the UK for a hundred years. Bats are protected species in many countries, and if roosting colonies are unwelcome, the best course of action would seem to be careful physical bat-proofing of roof spaces and other suitable roosting voids before colonization occurs. Even outdoor-roosting bats can pose problems if near to habitation or amenities. Bats in parks in Sydney, Australia, are subject to control experiments in which sounds projected through loudspeakers are used in an attempt to prevent concentrations of roosting bats from becoming too large.

DAMAGE CAUSED BY PLANT GROWTH

The colonization of buildings by plants is often similar to the colonization of newly exposed rock in nature. Plants exploit and help to create microenvironments suitable for plant growth, and, if left undisturbed, a succession of plants of increasing diversity and size takes place. The surface of stonework may be first colonized by microorganisms and small algae if conditions are suitable (enough moisture – which may be related to stone texture and porosity), followed by lichens and mosses, which hold water, thus aiding physical weathering. Such growths help mineral and organic debris to accumulate, and eventually a soil of sufficient depth and quality may be formed to allow the germination and growth of higher plants. In crevices in stonework there may be no real or suspected succession; once wind-blown soil and other debris have lodged there, higher plants may be able to grow straight away.

The effects caused by these various types of plant growth are discussed in the following subsections.

Lichens

Lichens disfigure buildings, often in a less obvious way than algae. Many of the smaller forms grow as minute dark colonies, tightly adpressed to the substrate. To the untrained eye, these colonies appear as specks of dirt, and indeed for many years, what was lichen growth on many buildings in towns and cities was regarded as soot or general urban dirt.

The larger forms of lichen, which grow in cleaner air than that found in industrial or urban centres, can, albeit slowly, play a part in the erosion of stonework. There are probably several mechanisms involved. Lichens are able to withstand desiccation and rehydration over a much greater range than other plants. This enables them to survive in extreme environments which could not be tolerated by other plants. The holding of water may, in itself, help in the physical weathering of stone caused by frost action in cold climates. Another physical mechanism, probably related to expansion and contraction of the colony with changes in water content, and also with actual growth, is the plucking of particles from the stone surface by adhering or penetrating filaments of the lichen thallus. The lichen colony may also remove minerals from the surface of the substrate, eventually causing exfoliation of altered stone, leaving a fresh surface to be recolonized and repeat the process. There may also be slow changes in some stone caused by the leaching of acidic waste products from the plant.

Mosses

Mosses are able to penetrate some types of stone with their rhizoids, and, as they are usually more substantial plants than lichens, play a larger role in trapping debris and building up soil. Once detached, by heavy rain or excessive drying, moss clumps can and frequently do block drains and gutters.

Both lichens and mosses growing on buildings can give a pleasant and mellowing effect; indeed, to tone down the stark whiteness of modern farm buildings, spraying with animal slurry to lower the pH and provide nutrients to hasten such growths has been advocated. The problems come later, and a balance has to be struck between aesthetic improvement and the long-term effects of stone deterioration. There is really no case for allowing plant growth of any kind to develop on buildings or monuments of historic or cultural importance.

Higher plants

Higher plants can colonize the actual structure of buildings if sufficient root anchorage is available, either in accumulations of soil on flat surfaces or in cracks and crevices (Figure 4.12). The species are diverse, ranging from small weed species through to woody shrubs and trees. The roots of woody perennials such as sycamore or yew saplings may penetrate extensively through stone and brickwork and induce cracking of the structure. Once

Figure 4.12. Seedlings of higher plants growing in a crack in a wall in Porto Alegre, Brazil. Root pressure will soon enlarge the crack. Photo: Dr D. Allsopp.

this stage has been reached, removal of the plants and restoration of the structure may involve complete reconstruction of affected sections.

Control of plant growth on buildings

Physical maintenance

Gaps that are due to missing mortar, bricks, or stones should be made good. Drains and gutters should be kept clear and the building generally kept free of dirt and litter. Frequent inspections should be made, and any plants which start to grow should be removed as soon as possible. Although these actions are obvious, they can be difficult and costly to perform, especially on large buildings which may need scaffolding to enable such work to be carried out. Some ancient monuments in the UK may be inspected only at intervals several years apart. In some of the poorer developing countries there may be no inspection or care at all at some historic sites.

Use of biocides

A range of washes can be used, including solutions of bleach (sodium hypochlorite), phenolic compounds, and quaternary ammonium compounds. Combinations of quaternary ammonium compounds with tributyl tin oxide or potassium orthophenyl phenol have proved particularly effective. Environmental considerations (pollution caused by run-off) and length of effectiveness have to be considered, as well as safety legislation. Many such treatments are still the subject of study, and many have already been banned. Porous items sufficiently small to be treated (e.g., statues) can be pressure treated with waterproofing compounds containing biocides.

It is difficult to 'rot-proof' stone in the same way as for wood. The biocides do not penetrate in the same way, and even on the more porous stone the compounds do not adhere and may leach out. Attempts at more long-term protection have been made by use of metallic copper strips high on buildings to provide a constant source of toxic copper ions, carried over the lower stone by the action of rain. The problems here are achieving an even and effective spread of copper without leaving the structure stained by green copper carbonate. Even where microbial growth is inhibited, the structure of the stone is not necessarily protected. Figure 4.13 shows a limestone monument in London, where copper leaching from bronze statues prevented colonization of the stone but not spalling of the surface. It was suggested that copper may react with the stone to produce an impermeable surface layer, causing salt build-up and eventual exfoliation.

Plants near buildings

Climbing plants

Creepers and vines are often well adapted to colonize buildings and may well be encouraged to do so for aesthetic reasons. In early and well-tended stages, there may be few problems, but well-established and neglected growth can give rise to a variety of effects. Such growth obviously provides harbourage for other organisms such as birds and insects, which may cause problems in their own right. Entry of rodents into upper-story

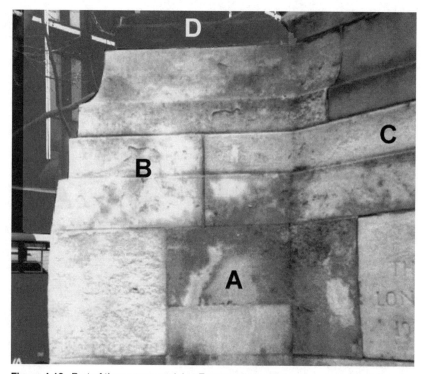

Figure 4.13. Part of the war memorial at Euston railway station, London, showing green discoloration by leaching copper salts and spalling. A, green discoloured stone; B, spalled area; C, stone with various degrees of grey discolouration; D, base of bronze statue. Photo: Dr Christine Gaylarde.

windows and vents or roof spaces is made much easier, and inspection of the building structure or fittings such as pipework and external cables can be made extremely difficult. The actual structure, whether of stone, brick, or wood, may be penetrated by suckers or shoots, causing physical damage. On small outbuildings, unchecked growth of creepers allowed to cover roofs can impair drainage, penetrate soft roofing materials (e.g., tarred felt), and put considerable strain on the roof. Under these circumstances, fungal decay of susceptible material in the roof and other parts of the building may be accelerated.

The crucial factor in these cases is correct maintenance of growth, with adequate annual inspection and trimming, and removal of growth from areas where problems may occur in future, such as drain pipes (Figure 4.14), gutters, and vents.

Figure 4.14. Wisteria stem entwining and crushing a zinc drainpipe. Plants on buildings should not be allowed to grow on utility fittings, only on plant supports provided for the purpose. Photo: Dr D. Allsopp.

Trees

Many trees are planted or allowed to grow near buildings with little thought as to the problems they may cause when they near maturity. Even removal of self-seeded saplings (e.g., sycamore) can be a considerable problem after only a few years' growth. The spreading branches of trees may grow outwards to touch buildings and, importantly, windows and cause damage as the tree grows further or sways in high winds.

The main problems, however, are caused by tree roots. It should be appreciated that tree roots can travel outwards from the base of the trunk at least as far as the tree branches and often much farther. Tree roots can penetrate and block drains or water supply pipes and do considerable damage to the foundations of buildings. Large breaks in drains may lead to subsoil erosion caused by the water released. Indirect damage may also be caused by the roots absorbing water from the soil, leading to excessive soil shrinkage, especially in clay soils in which resorption of water may not restore the original conditions.

The Royal Botanic Gardens at Kew, UK, have been asked many times to identify tree roots possibly responsible for such problems. A seminar held at Kew in 1976, 'Tree roots and damage to buildings', showed the need for more and better data. The Kew Tree Root Survey was accelerated and a book was published giving data on commonly planted trees (Cutler and Richardson, 1981).

Trees near buildings, roads, or other structures should be planted with care and consideration and with thought to their final size and root spread, bearing in mind the massive damage which may be caused in later years if good management and a cycle of removal and replacement is neglected.

HISTORIC AND CULTURAL BUILDINGS

The same organisms and types of biodeterioration as previously described also apply to historic buildings of cultural heritage, but, in this case, the losses are not merely economic. Most such constructions have already suffered deterioration, and the aim is to minimize decay while at the same time maintaining the appearance and important historical features of the building. Generally it is not possible to use surface coatings to this end, although painted buildings obviously continue to be repainted regularly. Curators and conservators responsible for the upkeep of historic buildings are, of necessity, very concerned with preservation and history, and this makes them loath to consider the use of biocides, which, in their eyes (and not incorrectly, if the biocides are thoughtlessly used) lead to long-term damage to the building itself and potentially to the environment. Even the use of a modern paint is anathema to some purists, who would like to see the entire structure of the building retained in its original form. Nevertheless, historic buildings *are* repainted and biocides are, occasionally, used, most frequently in cleaning the surface before painting.

The oldest constructions are not usually painted, either because this was not part of the original builders' plans or because the paint has been removed long ago. In these cases, protection of the surface is normally ensured by normal maintenance and cleaning. Only relatively recently has money been put into research on biodeterioration of cultural heritage by national governments, but there are currently a number of European and Latin American government funded projects in this area, many of which involve the use of techniques of molecular biology (see Chapter 5).

The use of DNA analysis has led to some unexpected results on the microbial populations present on historic buildings. Members of the *Archaea* (so called because their properties indicate that they are an ancient group of organisms) have been detected for the first time on buildings. This group of extremophilic bacteria, members of which grow at high temperatures, low water activities, and in the absence of air, are, in fact, not so out of place on building surfaces. The conditions on an external wall, for instance, vary enormously throughout the year and may include extreme desiccation and, in the tropics, very high temperatures. Another interesting result of the application of molecular biological techniques is the fact that a previously little considered member of the actinomycetes, *Geodermatophilus*, is a frequent colonizer of ancient stone surfaces.

At the time of writing, molecular methods have only recently been successfully applied to cyanobacteria on historic buildings; there is work in progress in a number of countries, and, by the time this book is published, considerable information will doubtless be available in the scientific literature. The current literature is somewhat confusing, as a bacteriological and several botanical taxonomic schemes are used for this group. DNA analyses of aquatic cyanobacteria have shown that these taxonomic schemes will have to be modified when the databases are complete, but for the moment we use here the bacteriological classification as given in Bergey (1994), although this was revised in 2001.

Members of all the five subgroups (now subsections) of cyanobacteria have been reported on ancient buildings. There is little doubt that the major cyanobacterial biomass (and frequently the major microbial biomass) on ancient calcareous stone buildings belongs to subgroup 2, the *Pleurocapsales* (Figure 4.15). In fact, the single-celled and colonial cyanobacteria, belonging to subgroups 1 and 2, seem to be more common on stone surfaces in general. On more porous cement-based surfaces, however, the number of filamentous types increases. These (subgroups 3–5) are also more easily isolated in culture media and therefore may sometimes be falsely reported as the main colonizers if the original biofilm is not inspected directly. Figure 4.16 shows a *Scytonema* sp. isolated from a painted church in Porto Alegre, Brazil. Although not the major biomass on the surface, it was most easily isolated in pure culture because of its speed of growth and its ability to move away from other organisms in the sample.

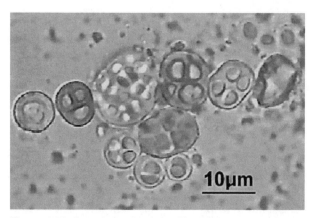

Figure 4.15. Cyanobacteria of the *Pleurocapsa* group, frequently the major biomass found on calcareous historic buildings. Photo: Dr Christine Gaylarde.

The opening of properties of cultural interest to the public puts them at a greater risk of biodeterioration. Unless sufficient funds are available to ensure that conditions are inimical to growth of organisms, exhibits on show will be subject to high levels of microbial (and nutrient) contamination brought in by the public and suitable levels of illumination for the growth

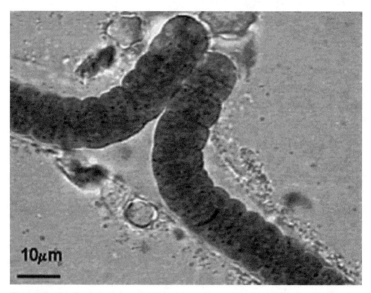

Figure 4.16. *Scytonema* sp. isolated from the outer discoloured wall of a church in Porto Alegre, Brazil. Photo: Dr Christine Gaylarde.

Figure 4.17. Plants growing on a stone building in Ouro Preto, Brazil. Photo: Dr D. Allsopp.

of phototrophs. Only constant vigilance and rigorous maintenance will be successful in preserving the materials.

Lower and higher plants grow on neglected historic buildings, as on modern ones, and the same precautions should be taken. In the tropics and sub-tropics, epiphytes are common on ledges of buildings (Figure 4.17), but then epiphytic plants can also be found growing on electric wires above the street, giving some idea of the extreme conditions of humidity prevailing!

In tropical and sub-tropical countries, the organisms most widely recognized by the general public, as well as by conservators, as damaging to historic (as well as modern) buildings are the termites. Any wooden structure is at risk unless physically protected. Because termites may enter through the soil or the air, this is a daunting task, and constant vigilance is necessary to deal with new infestations. Termites may be introduced into temperate regions and have now established themselves as a major threat, for example, in South-West France. Non-invasive methods of control, apart from total isolation of the buildings, are, unfortunately, impossible, and biocides, traditionally pyrethroids and organophosphorus compounds, are used. A number of alternative treatments are now under consideration. Pheromones, juvenile hormone analogues, and the essential oils of some plants have been considered, without, so far, apparent success. Inhibitors of chitin synthesis (chitin is the main structural component of insect exoskeletons) such as hexaflumaron are possible options as control agents.

Bird control is another important aspect of the conservation of historic buildings. Although the most acceptable method is to prevent birds from landing on the building, modification of the structure of important cultural properties is not allowed and less effective remedies, such as the use of 'scare-birds' (wood, metal, or plastic representations of birds of prey) may be considered. Bait doped with avian contraceptives has also been suggested as a long-term control strategy, but has not yet proved effective!

Transport systems

In this section, the obvious major problem of higher plants acting as deteriogens is considered. Railways, waterways, and roads are usually built at the expense of a natural flora, the climax of which, in most regions of the world, consists of communities of higher plants. If the systems are not maintained, there is a natural tendency for reinvasion to occur from the margins by vegetative spread or by seed.

RAILWAYS

Problems caused by plant growth can be seen in stockyards, track beds, and their surrounds. It should be noted that problems in the tropics (especially in humid areas) can be much greater than in the UK and other temperate regions. In stockyards, obstruction is a major problem; vision impaired can result in increased risk of accidents during shunting operations. As stockyards are obviously associated with storage, involving buildings and communication systems, the potential fire hazard may also be great, as is the problem of harbourage of vermin and even human pilferers. On the track itself, plants may cause visual and physical obstruction of both vehicles and track mechanisms. They increase the fire hazard on minimally manned track, cause changes in the movement and drainage of ballast supporting the rails on ties, and wheelslip. Safety of workers and ease of inspection of track and associated signal systems may also be impaired.

To combat the problem of undesirable plant growth, many railway systems use specially adapted vehicles fitted with sprayers to apply herbicides to the track and surrounds. The cost of these special coaches and the chemicals used is obviously considerable; a balance has to be struck between the cost of control and the financial benefits achieved.

As an aside, although steps are taken to discourage plant growth on tracks, some work in the UK has taken place to determine how best to encourage bacterial growth. On non-electrified lines where diesel locomotives are used, oil from engines can foul the ballast in areas where locomotives stop. This creates problems for the workforce. To improve the situation, extra nutrients and water-holding materials have been added to help microorganisms break down the oil.

ROADS

To avoid erosion on modern road edges, which often have embankments or cuttings with sloped sides to avoid steep gradients, some vegetation is usually desirable. Maintenance, varying from simple mowing to landscape gardening, is necessary to prevent plants becoming a nuisance. Visual and physical obstruction are obvious problems, as on railways. Even in towns and cities, deliberately planted decorative trees can cause problems. Overgrown branches can obstruct high vehicles and foul buildings and overhead lines, especially when moving in high winds; sticky resins may drip onto pavements and cars (one very famous car manufacturer advised its customers not to park under lime trees in spring); and where trees are used as roosts by birds, droppings can become a nuisance.

Other problems caused by plants, alive or as dead plant litter, include the blocking of drainage ditches, gratings, and culverts and the disruption of road surfaces. A dry summer, such as that experienced in the UK in 1976, highlights the hazards of fire on the surroundings of high-speed roads, and dense plant growth along road margins can cause uneven drying of soil, leading to uneven road surfaces.

The problems of drainage, surface disruption, communication system fouling, and animal and bird harbourage, which occur along roads, also apply to airfields, where the consequences are much more serious.

WATERWAYS

Plants in waterways, both floating and rooted, can give rise to problems by increasing water loss by transpiration, acting as silt traps, causing obstructions to water flow and navigation, blocking pipelines, complicating the disposal of pollutants, and reducing the amenity value.

In tropical and sub-tropical areas, the water hyacinth (*Eichornia crassipes*), a floating plant, and the alligator weed (*Althernanthera philoxeroides*), a plant which may also root in shallow water, cause problems by their vigorous growth, and a considerable amount of money is spent on controlling such plants. The growth of the water hyacinth is so vigorous that it has been used experimentally to remove excess nutrients from polluted waters and then harvested to be used as animal feed; heavy metals have been removed from polluted waters in the same way.

Control measures

Physical control
The pulling or cutting of weeds by hand or with manual implements is a traditional aspect of cultivation of land and tends to be slow and laborious. In many areas of the tropics, however, it may be the only method available. Burning is also a traditional method, but there may be difficulties in fire control, more important with transport systems than with agricultural land, and fire may stimulate the growth of different plant species and merely change the nature of the problem.

Mechanical control
Modern power-driven mowing machines are available for use on verges of roads, railway, and waterways, and heavy-duty versions can deal with tree boles up to about 25 cm in diameter. Mechanical techniques are also available for use with aquatic weeds, the submerged cutting heads being powered by a motor mounted on a boat or on a vehicle on the bank.

The major drawback to simple mechanical control is that many plants respond to cutting by vigorous regrowth, and permanent control by such methods is almost impossible.

Biological control

Such methods often centre on the use of insects, and, as with all cases of biological control, an in-depth picture of possible consequences must be obtained before treatment. Insects may spread over a wider area than intended, and even if specific to a limited group of plants, such plants may not be considered as deteriogens in other areas. Transport systems can be considered as 'narrow and long', and the insect-based systems of biological control, even if used successfully in an agricultural 'area' sense, may not be suitable. The most promising application of biological control of plants affecting transport systems may well be in waterways, where spread of organisms is much less likely. A wide range of potentially useful organisms has been suggested, including microorganisms, aquatic insects, snails, fish, and even vertebrates such as the manatee.

Chemical control

Manufacture and use of chemicals to control growth of vegetation comprise an industry in their own right, such is their widespread use. Initially, the development of herbicides was aimed at agricultural weeds, and these were then used to control plants in other situations. However, the industry is now of such an extent that different herbicides are developed for a variety of situations. Toxic herbicides may be classified into three main types according to their action, and, because of this, may have different applications in controlling biodeterioration.

Contact herbicides have a caustic or corrosive action on plant tissue; many of the early herbicides were of this type.

Systemic herbicides enter the plant through the leaves, whereupon they are translocated throughout the entire plant, where their toxic effects ultimately cause the destruction of the plant.

Soil sterilants are added to the soil and have the effect of rendering the soil toxic to plants. The soil sterilant is taken up by the roots, translocated, and hence destroys the plant; thus it is essentially a systemic herbicide.

A further classification is on the basis of range of action. 'Total' herbicides control most plant species, and selective herbicides usually control either broad-leaved or narrow-leaved plants. Another group of herbicides, the growth regulators or plant hormones, in small doses inhibits various aspects of the plants' development and in high doses is lethal.

The use of chemicals to control plant growth, either on transport systems or buildings, or indeed in any context, cannot be considered as a complete solution. Aspects of environmental pollution have to be considered, especially in the cases of transport systems and buildings, in which there is usually the close presence of humans, their animals, and crops. Even if plants are killed, they do not disappear immediately, and the drying residues may for some time still cause problems of a fouling, fire hazard, or aesthetic nature.

TRANSPORTATION – AIRCRAFT AND SHIPS

The following are main problem areas:

1. Aircraft
   a. Bird strikes
   b. Pest transmission
   c. Fuel problems (see the subsection on fuel and lubricant problems)
2. Ships
   a. Hull fouling
   b. Fuel and lubricant problems
   c. Cargo deterioration

Aircraft

Bird strikes

There are two main areas of concern, large flocks of birds on and around airports and birds singly or in flocks flying at altitude.

An aircraft is at its most vulnerable during landing and take-off; speed is critical, the flight-deck workload is greatest, and options on manoeuvre are extremely limited. The main problem here is the ingestion of birds into jet engine intakes. Small birds may cause minimal damage if ingested singly, but ingestion of large flocks can cause considerable damage and result in engine failure. Birds, of course, do not deliberately fly into aircraft, but tend to move out of the way of apparent danger; however, this is not an infallible principle and bird strikes do occur. With the advent of small 'corporate' or private business aircraft that use pure jet engines the problem may increase. New-generation jet engines on large airliners are much quieter than previously, and there is evidence that birds have much less warning of the approach of such aircraft.

Many methods of bird control have been attempted at airports. Control of the environment has been shown to be very important. The siting of municipal garbage dumps next to airports encourages large flocks of birds, especially gulls. Vegetation between runways can harbour birds, but short mown grass is not unattractive either. Bird scaring, by pyrotechnics, lights, recorded bird alarm calls, and even trained hawks, has been used.

Potential bird hazards may also be present at high altitudes. Bar-headed geese (*Anser indicus*) can fly as high as 10 000 m, where the temperature is −55 °C. These most efficient birds possess special blood cells to aid their high-altitude performance, as they migrate from high Central Asia over the Himalayas to Pakistan and North India, riding jet-stream winds of 200 miles per hour!

At altitude, when an aircraft is flying at 500 kn or more, a high-flying bird such as a goose, eagle, or vulture exhibits a remarkable biological property from the pilot's point of view. It becomes invisible. The chance of avoiding a head-on bird strike under these conditions is practically impossible. All that can be done is to ensure the airframe is strong enough to withstand a head-on strike of a bird around 2 kg in weight at the speeds used below 5500 m (18 000 ft). Such design considerations give the best practical solutions to the problems likely to be encountered. Most birds are below 2 kg in weight, and most strikes take place at lower altitudes. An aircraft proofed against the heavier birds (such as geese or swans) at full high-altitude operating speed (500 kn or more) would be much too heavy to be practical. A single-engine shutdown at cruising speed is much less critical than at landing or take-off, and the subsequent landing on fewer engines can be planned with more time to spare.

We now know more about bird migration routes, and with this knowledge and the use of radar, at least the well-known seasonal flocks of birds can be avoided by regular scheduled flights.

Pest transmission

The modern airliner is a crowded, heated, restaurant–bar–lounge and is subject to most of the pest problems of fast-food outlets. Pests can enter with passengers, luggage, and cargo, with catering supplies, and from the air when the aircraft is on the ground. Perhaps more significantly, the vast distances covered by aircraft in short periods of time give them a potential role as vectors of pests or of disease organisms carried by them. To prevent all these problems, aircraft are fumigated regularly and particularly after any insect or rodent pests have been spotted. On some routes, airline staff may use handheld aerosols of insecticide before each flight. For major fumigations methyl bromide is normally used, but the work is costly, especially in time. At London's Heathrow airport, pest-control companies on an approved list have only a few hours to respond if called out. The actual job takes about 8 hours, often overnight, and requires the presence of engineers at the start and finish to turn on power and to open and close the aircraft hold, hatches, and doors. Pest-control staff, who have to be present by the aircraft throughout to prevent accidental entry, check it is clear of fumigant by using detector meters (and wearing respirators!) at the end of the operation. Because of the phase-out of methyl bromide over the next few years, alternative treatments such as carbon dioxide are being evaluated. The very sensitive environment of electronics means that many reactive chemicals are ruled out.

Ships

Hull fouling

Some algae and sessile animals such as barnacles and sponges are able to attach themselves to submerged portions of ships' hulls at early stages in their life cycles, and there develop and grow. Most exposed surfaces in the first few metres of the sea are subject to the same process of colonization; for example, rocky shores, piers, floating pontoons, and oil rigs show similar patterns. In the case of ships, the increased water resistance that is due to such fouling leads to a slowing of the ship, resulting in increased fuel consumption to compensate. The higher the performance the ship, the more marked the effect. Details of effects on naval ships are not easily available, but it is estimated that the fuel consumption of a fast warship could increase by 50% in as little as 6 months in waters where rapid fouling occurs.

   Protection of ships' hulls against fouling (and against the marine boring animals which attack wooden ships) has a long history. Copper sheeting was once used, and a range of toxic coatings have been employed over the years. Today, the usual remedy is the use of antifouling paint containing a toxic chemical. The life of such coatings is very important, as the cost of dry docking a ship to renew the antifouling is high. The latest coatings are self-polishing; the paint film is both toxic and unstable and wears away at a controlled rate. Any beginnings of fouling growth are swept away before they become significant and a fresh toxic surface is exposed. However, to coat a hull in such a way as to give an even length of life is a problem, as wear in some areas (e.g., around the bows) is greater than in others. There is also a potential for environmental pollution because of the release of the toxin into the water, and non-toxic alternatives are being sought. One possibility is biological control. When an object is placed in the sea, it begins to develop a film of bacteria very rapidly (within hours). In some circumstances this film may be of significance in the subsequent settlement and growth of fouling organisms; some strains of bacteria appear to enhance settlement of algal spores and their germination; other strains appear antagonistic. Current work is looking at ways in which the film may be manipulated to discourage subsequent fouling.

Fuel and lubricant problems

Marine diesel engines are usually robust and well-tried units and will tolerate some variation in fuel quality. This is not true of gas-turbine engines. Marine gas turbines are basically the jet engines used in aircraft, and their use is normally confined to naval vessels. The fuel must be free of seawater, and a system of fuel filters, water coalescers, and centrifuges is used to remove water and clean the fuel before it is fed to the engines. If the system has to cope with large amounts of fungal growth, fuel lines and the processing system can become blocked and engines can fail. Fungi can also develop on the filter elements, contributing to premature failure. The organisms involved are basically the same as those found in aircraft fuel tanks, for the fuel is very similar. The problem may be increased by use of seawater-displaced fuel tanks. To keep the ship in trim (i.e., a correct weight distribution) as fuel is used, it is replaced with seawater; the fuel floats on top. This ensures a source of extra nutrients and a permanent and large fuel–water interface where organisms can grow.

The main control method has been the addition of a biocide (usually an organoboron compound). The difficulty arises in maintaining an even and effective dose level, especially under operational conditions. Future ship design may employ some means of pasteurizing fuel by using the abundant waste heat found on ships, as well as using biocides.

Although the fuel of marine diesel engines usually presents few practical problems, the lubricants employed can be subject to microbial attack. The large pistons in these engines are water cooled, and there is usually some degree of water contamination of the lubricating oils. The temperatures attained are much lower than those in car engines where self-sterilization would take place. At a temperature of only 40 °C, despite various fuel renovation systems used, bacterial and fungal growth can break down the lubricant, leading to rapid wear on the bearings. More efficient water

separation and the use of heat pasteurization systems could also be employed to good effect here.

Cargo deterioration

This is a huge topic and can be considered only very briefly here. It encompasses most of the field of stored product biology. Some cargoes are
so obviously perishable that special ships have been built to carry them,
classic examples being refrigerated ships to carry meat from New Zealand
to the UK and ships with cooled holds to transport bananas, allowing
their timed ripening to occur en route. Foodstuffs are generally the most
perishable cargoes, and the shipping world has developed a legal system
rather than a scientific one to deal with problems. It would seem that insurance is cheaper than detailed inspection before shipment and better
environmental care en route. Scientific investigation of spoiled cargoes
is usually centred on apportioning blame rather than in improving the
technology of shipment.

New practices can also pose problems. Wood has often been carried as
deck cargo. If the wood is machined, rather than logs or huge timbers, it is
obviously best protected from salt spray. Heavy-duty black polythene has
been used for this, and it was found that daytime heating by the sun and
cooling at night led to water migration from the timber, condensation of
water inside the package, and massive growth of fungi, leading to staining
of the wood. A biocide treatment would be used to prevent this.

Time is an important consideration with sea transport. A normal voy-
age may use up a considerable fraction of the 'shelf life' of perishable or
semi-perishable goods. Delays that are due to weather, political upheaval,
customs, and strikes at ports may make all the difference. The advent of
roll-on roll-off (RORO) ferries for shorter journeys and the increasing use
of containerized transport have led to improvements in many areas.

MUSEUMS

Museums are a special category of buildings, with respect to risks of
damage caused by organisms. Many museums are old. They may have
been purposely built for display of objects, but are rarely purposely built
for controlled environments and minimizing biological risks. Most have
been modified over a long period in regard to heating, lighting, ventilation, access, and power supply. Most are too small, especially in regard to
the storage of materials not on display. There is often a large amount of

material housed in rooms not on view to the public, sometimes for the use of scholars, but often merely because of lack of space and funds to enable proper display.

The objects in museums also fall into a special category. The items are usually rare, if not unique. They are often composite in nature and may be fragile. Standard practices in the preservation of materials do not hold good in the museum environment. Removal of fragments for screening of biocides is usually not possible. Many biocides are ruled out because of colour changes, toxicity, and the long-term risks of chemical damage. The craft of conservation work dictates that any treatment should be reversible. This is ideal in theory, but a compromise may have to be reached in practice. Members of a museum staff need to think of objects preserved in perpetuity and not merely protected for a finite working life. The ideal museum environment would be one of total environmental control, with each object sealed from the outside atmosphere, supplied with filtered air at a safe and suitable humidity, kept at the lowest temperature possible bearing in mind the physical nature of the object, and illuminated at low light levels. However, the demands on space, money, and academic and public involvement mean that a compromise is inevitable.

On arrival, objects need to be conserved. The procedures are outside the scope of this book, but suffice to say they involve cleaning, repair, and stabilization of the object (including the elimination of any biological infestation) before it can go on display. Conservation should not be confused with restoration, the latter being an attempt to bring an object back to its original state, often with the addition of new parts and new surface finishes. Restoration of objects is a part of the antiques trade rather than conservation of museum objects.

Elimination of biological infestation may be dealt with by techniques which impart no lasting chemical residues, provided that subsequent handling, storage, or display conditions are not conducive to biological growth. Recent developments include the use of low temperatures ($-30\,^{\circ}$C for 3 days or $-18\,^{\circ}$C for 2 weeks), high temperatures at controlled humidity (52 $^{\circ}$C for 2 hs), nitrogen anoxia, and carbon dioxide fumigation. Many of the commercial formulations we could use to protect a wooden fence would also work on ancient wood carvings in museums, but the irreversible nature of the treatments and the lack of knowledge about long-term effects rule them out from the beginning. A water-based microemulsion formulation of permethrin has been developed for use on objects such as textiles and furniture, where other methods of disinfestation are not practical.

Good housekeeping and environmental control, which should be the basis of all biodeterioration prevention, are of paramount importance in the museum environment and cannot be overstressed. The following comparison of procedures for overcoming biodeterioration indicates the importance of applying correct and lasting treatments.

Ancient chair found in poor storage.	Old garden seat found in poor storage
Some mould growth and woodworm.	Some mould growth and woodworm.
Specialist assessment, carbon dioxide fumigation or anoxic treatment, careful cleaning, drying under controlled humidity. Sufficient repair to hold chair together. (Job perhaps over several years).	Dry out over a few days. Sand down with power tools. Replace all missing parts with new wood. Liquid insecticide for woodworm. Wood preserver for mould. New varnish or surface finish. (Job perhaps over 1–2 weeks.)
Display in air-conditioned showcase, fitted with environmental monitoring devices.	Use outdoors.
Permanent exhibit. Regular checks for infestation	Eventual repeat of problem. Need for eventual replacement.

REFERENCES AND SUGGESTED READING

ADAS Pest Manual. (1999). Agricultural Development and Advisory Service (ADAS), Guildford, England.

Allsopp, D. and Gaylarde, C. C. (2002). *Heritage Biocare; training course notes CD.* Archetype Publications, London.

Allsopp, D. and Drayton, I. D. R. (1975). The higher plants as deteriogens. In *Proceedings of the Third International Biodegradation Symposium*, Sharpley, J. M., and Kaplan, A. M. (Eds.). Applied Science Publishers, London, pp. 364–75.

Baynes-Cope, A. D. (1989). *Caring for Books and Documents*, 2nd ed. British Museum Publications, London.

Bech-Andersen, J. (1995). *The Dry Rot Fungus and Other Fungi in Houses.* Hussvamp Laboratoriets Forlag, Holte, Denmark.

Bletchly, J. D. (1967). *Insect and Marine Borer Damage to Timber and Woodwork – Recognition, Prevention and Eradication.* Her Majesty's Stationery Office, London.

Boone, D. R., Castenholz, R. W., and Garrity, G. M. (2001). *Bergey's Manual of Systematic Bacteriology*, 2nd ed. Vol. 1. Springer, New York.

Buckle, A. P. and Smith, R. H. (Eds.). (1994). *Rodent Pests and Their Control.* CAB International, Wallingford, Oxon, U.K.

Burger, J. (1983). Bird control at airports. *Environ. Conserv.*, **10**, 115–24.

Brooks, F. T. and Hansford, C. G. (1922). Mould growths upon cold-store meat. *Trans. Br. Mycol. Soc*, **8**, 113–42.

Callow, E. M. and Callow, J. A. (2002). Marine biofouling: A sticky problem. *Biologist*, **49**, 10–14.

Coggins, C. R. (1980). *Decay of Timber in Buildings*. Rentokil Library (Rentokil Library is found on-line through a number of search engines, such as www.google.com).

Cornelius, M. L., Grace, J. K. and Yates, J. R. (1997). Toxicity of monoterpenoids and other natural products to the Formosan subterranean termite (Isoptera: Rhinotemitidae). *J. Econ. Entomol.* **90**, 320–5.

Cornwell, P. B. (1968). *The Cockroach*. Rentokil Library, Vol. 1.

Cornwell, P. B. (1979). *Pest Control in Buildings*, 2nd ed. Rentokil Library.

Cornwell, P. B. (1976). *The Cockroach*. Rentokil Library, Vol. 2.

Cowan, D. P. and Feare, C. J. (Eds.). (1999). *Advances in Vertebrate Pest Management*. Fillander-Verlag, Fürth, Germany.

Crispim, C. A., Gaylarde, P. M., and Gaylarde, C. C. (2002). Algal and cyanobacterial biofilms on calcareous historic buildings. *Curr. Microbiol.*, **46**, 79–82.

Cutler, D. F. and Richardson, I. B. K. (1981). *Tree Roots and Buildings*. The Construction Press, Lancaster, England.

Davis, J. G. (1951). The effect of cold on micro-organisms in relation to dairying. *Proc. Soc. Appl. Bacteriol.*, **14**, 216–42.

Davis, R. A. (1961). *Control of Rats and Mice*. Bulletin 181, MAFF. Her Majesty's Stationery Office, London.

Edwards, R. and Mill, A. E. (1986). *Termites in Buildings*. Rentokil Library.

Flannigan, B. (1991). Deteriogenic micro-organisms in houses as a hazard to respiratory health. *IBS Symp. Proc.*, **8**, 220–33.

Flannigan, B., Samson, R. A., and Miller, J. D. (Eds.). (2001). *Microorganisms in Home and Indoor Work Environments*. Taylor & Francis, London.

Florian, M. L. E. (1997). *Heritage Eaters – Insects and Fungi in Heritage Collections*. James and James, London.

Florian, M. L. E. (2002). *Fungal Facts – Solving Fungal Problems in Heritage Collections*. Archetype Publications, London.

Garcia de Miguel, J. M., Sánchez-Castillo, L., Ortega-Calvo, J. J., Gil, J. A., and Saiz-Jimenez, C. (1995). Deterioration of building materials from the Great Jaguar Pyramid at Tikal, Guatemala. *Build. Environ.*, **30**, 591–8.

Gaylarde, C. C., Ribas Silva, M., and Warscheid, T. (2003). Microbial impact on building materials: An overview. *Mater. Struct.*, **36**, 342–52.

Gaylarde, P. M. and Gaylarde, C. C. (1998). A rapid method for the detection of algae and cyanobacteria on the external surfaces of buildings. In *Proceedings Third Latin American Biodegradation and Biodeterioration Symposium*, Gaylarde, C. C., Barbosa, T. C., and Gabilan, H. N.(Eds.). The Phycological Society, England, Paper No. 37.

Gaylarde, P. M., Gaylarde, C. C., Guiamet, P. S., Gomez de Saravia, S. G., and Videla, H. A. (2001). Biodeterioration of Mayan Buildings at Uxmal and Tulum, Mexico. *Biofouling*, **17**, 41–5.

Gill, C. O. and Lowry, P. D. (1982). Growth at sub-zero temperatures of black spot fungi from meat. *J. Appl. Bacteriol.* **52**, 245–50.

Guidelines on Assessment and Remediation of Fungi in Indoor Environments. New York City Department of Health. Bureau of Environmental and Occupational Disease Epidemiology.

Hickin, N. E. (1971). *Wood Preservation.* Rentokil Library.

Hickin, N. E. (1975). *The Insect Factor in Wood Decay,* 3rd ed. Rentokil Library.

Hickin, N. E. (1981). *The Woodworm Problem,* 3rd ed. Rentokil Library.

Hickin, N. E. (1981). *Household Insect Pests,* 2nd ed. Rentokil Library.

Holt, J. G., Krieg, N. R., Sneath, P. H. A., Staley, J. T., and Williams, S. T. (1994). Bergey's Manual of Determinative Bacteriology (9th Ed.). Williams and Wilkins, Baltimore.

Houghton, D. R. (1978). Marine fouling and offshore structures. *Ocean Manage.,* 4, 347–52.

Jackson, W. B. (Ed.). (1995). *Vertebrate Deteriogens.* Int. Biodeterior. Biodeg., **36**, special issue.

Killgerm Pest Control Manual. (2001). Killgerm Group Limited, Ossett, West Yorkshire, England.

de Lelis, A. T. et. al. (2001). *Biodeterioracao de Madeiras em Edificacoes.* Instituto de Pesquisas Tecnologicas, Sao Paulo, Brazil.

Linnie, M. J. (1996). Integrated pest management: A proposed strategy for natural history museums. *Mus. Manage. and Curator.* **15**, 2.

Lucas, C. M. (1982). *Hygiene in Buildings.* Rentokil Library.

Meehan, A. P. (1984). *Rats and Mice.* Rentokil Library.

Meyer, A. N. (1998). *Rodents.* World Health Organisation, Geneva.

Mourier, H., Winding, O., and Sunesen, E. (1977). *Collins Guide to Wild Life in House and Home.* Collins, London.

Mueller, D. K. (1998). *Stored Product Protection . . . A period of transition.* Insects Ltd. Indianapolis, IN.

Ortega-Morales, O., Guezennec, J., Hernandez-Duque, G., Gaylarde, C. C., and Gaylarde, P. M. (2000). Phototrophic biofilms on ancient Mayan buildings in Yucatan, Mexico. *Curr. Microbiol.,* **40**, 81–85.

Pearson, C. (1993). *Building out pests.* AICCM Bulletin 19, 1&2.

Peterson, C. (2003). Insect repellents in urban settings. *Biologist,* **50**, 39–43.

Pinniger, D. (2001). *Pest Management in Museums, Archives and Historic Houses.* Archetype Publications, London.

Pinniger, D. B., Blyth, V., and Kingsley, H. (1998). Insect trapping: The key to pest management. In *Proceedings of the Third Nordic Symposium on Insect Pest Control in Museums.* 96–107.

Pinniger, D. B. and Child, R. E. (1996). Insecticides: Optimising their performance and targeting their use in museums. In *Proceedings of the Third International Conference on Biodeterioration of Cultural Property,* 190–9.

Proofing of Buildings Against Rats and Mice. (1965). MAFF Technical Bulletin No.12 Her Majesty's Stationery Office, London.

Purvis, W. (2000). *Lichens.* The Natural History Museum, London.

Richardson, B. A. (1980). *Remedial Treatment of Buildings.* The Construction Press, Lancaster, England.

Ridout, B. (2000). *Timber decay in buildings. The conservation approach to treatment.* E and F N Spon, London.

Robinson, W. H. (1996). *Urban Entomology.* Chapman and Hall, London.

Rossol, M. and Jessup, W. C. (1996). No magic bullets: Safe and ethical pest management strategies. *Mus. Manage. Curator.,* **15.**

Selwitz, C. and Maekawa, S. (1998). Inert gases in the control of museum insect pests. In *Research in Conservation.* The Getty Conservation Institute, United States of America. p. 107.

Sharma, R. N. and Raina, R. M. (1998). Evaluating chemicals for eco-friendly pest management – I: Terpenoids and fatty acids for building termites. *J. Sci. Ind. Res.,* **57,** 306–9.

Shirakawa, M. A., Gaylarde, C. C., Gaylarde, P. M., John, V., and Gambale, W. (2002). Fungal colonization and succession on newly painted buildings and the effect of biocide. *FEMS Microbiol. Ecol.,* **39,** 165–73.

Solfman, V. E. P. (1981). Birds and aviation. *Environ. Conserv.,* **8,** 45–51; Birds and aircraft. In *Research and Development Report: Mammals and Bird Pests.* Her Majesty's Stationery Office, London, pp. 43–4.

Strang, T. J. K. (1996). The effect of thermal methods of pest control on museum collections. In *Proceedings of the Third International Conference on Biodeterioration of Cultural Property,* 199–212.

Strang, T. J. K. (2001). Principles of heat disinfestation. In *Integrated Pest Management for Collections. Proceedings of 2001 – A Pest Odyssey,* Kingsley, H. (Ed.). James and James, London, pp. 114–29.

Sully, D., Man-Yee, L., and Swee Mun, L. (2001). A topical solution to a tropical problem. In Kingsley, H. (Ed.). *Integrated Pest Management for Collections. Proceedings of 2001 – A Pest Odyssey,* James and James, London, pp. 63–75.

Wee, Y. C. and Lee, K. B. (1980). Proliferation of algae on surfaces of buildings in Singapore. *Int. Biodeterior. Bull.,* **16,** 113–17.

Wilkinson, J. G. (1979). *Industrial Timber Preservation.* Rentokil Library.

Zycherman, L. A. and Schrock, J. R. (Eds.). (1988). *A Guide to Museum Pest Control.* American Institute for Conservation of Historic and Artistic Works, The Association of Systematics Collections, Washington DC.

5

Investigative biodeterioration

The prevention and control of a biodeterioration problem require a knowledge of the product composition, process and manufacturing details, supply chain information, and the application in order to be able to recommend the most equitable solution to all parties concerned. As the materials involved have a value, there may be the need to establish or apportion responsibility for a problem, and thus a forensic approach may be necessary in some cases.

The expertise in this area has, over the past 15 years, been taken up by the manufacturers and formulators of preservative products as part of their customer service package. They have developed the methodologies, built up databanks of experience, and have been proactive in the revision of national and international standards for the evaluation of preservative efficacy and materials durability for which microbial resistance is required. In a small number of areas, such as wood, stone, and food, government agencies still exist, and there are a small number of consultancies who can provide an independent view.

There are four elements to the investigation of a biodeterioration problem:
• Determining the cause
• Assessing likely control systems
• Instituting the most appropriate control system
• Monitoring the success (or otherwise)

THE PLANT AUDIT

Many biodeterioration problems have their origins in industrial processes, and it is impossible to appreciate the full range of factors involved unless a visit is made to the site of manufacture and a full survey of the situation is carried out. In this way information can be collected regarding the raw-materials storage and handling, the configuration of the process equipment, stages in the manufacturing process, and ultimate storage or distribution systems. Biodeterioration problems in the manufacturing industries are often complex, involving a number of interacting factors of an interdisciplinary nature. One of the major factors in this complex is the involvement of people who have scant knowledge of the topic and little or no training in biology. The investigator has therefore not only to be experienced and competent in the technical problems, but must also learn to be adept at the skills of detection, compromise, and diplomacy. Before any biodeterioration problem can be solved, it must be recognized. Non-microbial problems may be fairly easy to diagnose. Several centimetres chewed off the bottom of a door by rats, rodent teeth marks in wiring in a radio set, coupled with droppings, or armies of cockroaches on the march in a hospital ward are all fairly clear signs of a biological problem, and most people would recognize them as such. The situation with microorganisms is very different. By definition, these organisms are difficult to see and are often mixed in with the material being attacked. Where the effects are so gross as to give rise to some form of surface coating, the actual appearance is not necessarily different, to the untrained eye, from an accumulation of dirt or dried-up raw material. The microbial menu also provides a real stumbling block. The rat, mouse, or cockroach feeds itself on materials which are readily recognized as food, but how realistic is it to expect a member of a factory workforce to spot the activities of bacteria in a MWF, a paint, or a plastic? As a result, some preservative companies run company seminars to explain the activities of microorganisms, how to recognize their effects, and control measures.

An investigator called in to give assistance should talk to all the staff involved in the process, even if the initial contact, who will almost certainly be management, gives assurances that this is not necessary. The investigator must be diplomatic, as management is not always fully aware of the actual practices carried out at shop floor level, even if procedures are laid down in writing. The correct level of biocide addition in the factory is a good example of this problem. Raw-materials dosing is often automated in modern factories, with fail-safe systems which prevent the product from being manufactured until all the ingredients or pre-mixes have been incorporated. However, manual additions, particularly of small quantities (biocides may be added at as little as 0.1% of the total batch), may take place and then the question arises: Was the biocide added?

The next stage is to carry out a bioaudit of the plant. This involves taking samples of raw materials at various stages in their movement through the plant (don't forget the water tanks – often a repository for biofilm build-up!), identifying and sampling the critical stages in manufacture at which contamination could develop or be introduced, not forgetting any containers used to fill the product (filling lines are often sources of contamination). The investigator should come equipped with a suitable bag of 'tools' to cover all events. Sterile containers, a knife/scalpel, forceps, hand lens, swabs, a marker pen, and a means for sterilizing implements are some of the equipment required. It is generally useful for more than one investigator to be involved at this stage – one to keep the firm's representative busy while the other makes a more probing examination. The investigator should carry out the collection of samples in a methodical way, politely refusing offers of help and samples of unknown age and origin. It is important that the investigator should be satisfied that the samples collected are representative of the problem. This is stressed because it is all too easy to be talked into accepting a sample given in good faith in order to save time or the aggravation associated with getting to the site of sampling. The following short account will serve as an example.

A sample of river water was required for assessment of the microbiological susceptibility of a material in that environment. The sample was to be collected from the river at a site where there were building works, and arrangements had been made with the site engineer to gain access to the site during working hours. On arrival, the investigator was passed down the chain of command while one of the containers he had brought was pulled from his clutches and taken to a nearby standpipe. When he explained that river water was required, he was asked why he wanted dirty polluted water when there was plenty of clean drinking water available! Imagine the result if the containers had been sent to the site!

At the conclusion of the visit, the investigator may be asked to present a short verbal report to interested parties while still on the premises. There will almost certainly be a request for a short-term solution or alleviation to the problem. Only experience in solving problems can result in the ability to recommend such solutions. Good contact should be established with at least one member of the firm, who is able to act as a liaison, in order to expedite the solution to the problem. Once the visit has taken place, the investigating laboratory will examine the bioaudit samples for levels of contamination, differentiating bacterial and fungal growth. Identification of individual species is not to be recommended at this stage as it is both time consuming and costly and often not necessary in the long run. It will be appreciated that a rapid, but not unrealistic, response is necessary and an indication of time scale should be conveyed to the customer. Any report should be produced in a form which is clearly understandable to the non-biologist, and it should be borne in mind that the report may be passed on to third parties for examination. The authors among them have built up a bank of knowledge by which problems may be categorized into five types:

1. Single cause
2. Multiple cause
3. 'Red herring'
4. Inherently insoluble
5. Long-since-gone

Such categories have helped the authors to use case histories in biodeterioration teaching and indeed to devise selective methods for investigating the different categories once they have been defined. The categories, with appropriate case histories, are described in detail elsewhere, but are briefly subsequently reviewed.

The *single-cause* problem is usually quickly traced, and remedial action results in an effective solution.

The *multiple-cause* problem is the most commonly encountered and consists of a large number of interrelated factors. Each must be assessed before an effective regime of remedial action can be recommended. Monitoring of the effects of the action will be necessary.

Occasionally, the investigator must accept that a problem will turn out to be non-biological; this is the '*red herring*'.

Lack of information available to the investigator is often a contributory factor. The *inherently insoluble* problem arises from the inability to overcome environmental conditions and protect the material from medium-term deterioration; for example, the rapid leaching out of biocide under harsh conditions (such as repeated immersion in water or detergent solution) leaves an unprotected material which may subsequently be attacked by biodeteriogens.

In the final category, the *long-since-gone* problem, the case is not brought to the attention of the investigator until long after the initial attack has occurred, perhaps up to 10 years. This is usually the result of prolonged litigation or insurance claims in which one or both parties have not promptly sought biological advice. The product is then presented for assessment after the causative organisms have long since departed. As a result, conclusions are often circumstantial.

Assessment of control strategies

It is almost inevitable that in industrial product manufacturing there will be a need for improved plant hygiene and the use of an appropriate chemical preservative. The single reliance on either one of these systems is not a practical reality even in cosmetic preparation. We cannot expect a modern preservative to cope with the continued contamination from problems in the process for two reasons: preservative levels in products are limited by regulation and there is a risk of the development of microbial tolerance to the preservative. Good housekeeping and plant hygiene are the key to an integrated control system. The bioaudit will have identified the weaknesses in plant hygiene, and housekeeping measures can then be recommended.

A knowledge of the product composition and its application will help in the recommendation of suitable preservative systems. The type of contamination is also important as many preservatives are biased towards bacterial, fungal, or algal control. The investigating laboratory will then carry out a challenge or insult test to assess the ability of the product containing the candidate preservative to resist contamination of either single standard organisms (often specified for cosmetic products), a mixed inoculum from the laboratory, or organisms cultured from the plant (often known as 'house organisms'). A range of conditions can be employed to test the robustness of the preservative, and these are described in more detail later in this chapter.

Instituting the control system

The plant hygiene recommendations usually involve the introduction of cleaning procedures, water treatment, removal of dead ends, and the use of lids or head space disinfectant mists to control aerial contamination. In many situations it will be agreed that a works trial of the new preservative system should take place. This involves evaluating the preservative and/or the plant hygiene measures during manufacture of the product. Because this is a commercial operation, steps must be taken to make sure that the product does not leave the factory in an unfit condition. The purpose of this exercise is to establish that the recommended control strategies work in practice. A detailed bioaudit programme must be set up to assess samples from all the critical stages. If there are indications that the system is failing, then remedial actions must be put into place to limit the loss of product or enable it to be reworked.

Monitoring the effective system

Having established an improved control system, it is advisable that periodic bioaudits are performed by the preservative supplier or consulting laboratory so that early indications of potential problems are flagged up and remedial action can be initiated. A 'traffic-light' (green, orange, red alert) or numerical warning system might be used to prevent unnecessary panic at the first sign of a bacterial colony isolated from a sample. Depending on the process, a background or noise level of microbial contamination may be acceptable as long as it causes no adverse effect on the integrity of the product. An increase above this level would trigger a concern which would initiate a retest. If the retest confirms the first result, then a specified action would be initiated. If this brings the problem under control, then the normal monitoring would resume. The customer and investigating laboratory must determine between them a monitoring system which is appropriate to both parties.

DETECTION TECHNIQUES FOR BIODETERIOGENIC MICROORGANISMS

The standard methods of detection and identification of microorganisms are also used in biodeterioration. Textbooks and manuals which cover this area will be useful, especially if they consider organisms in the environment, rather than merely clinical specimens. Often, of course, it is not the identity or name of the organism that is important, but its particular deteriogenic activity, and some more recent tests aim simply at identifying enzymes or their specific products. An example is the hydrogenase test for SRB, which could be particularly relevant if the enzyme hydrogenase is important in biocorrosion.

Sampling

This is the most important phase of any microbiological analysis. If appropriate and representative samples are not available, even the most sensitive detection technique is worthless. One need consider only a silo full of stored grain which has to be tested for the presence of mycotoxins to realize the essential nature of this step. Only a handful of grains may be contaminated; these may all be present in the first sample, or may not be detected at all. Expert advice should be taken if a standard method is not laid down.

As we have seen throughout this book, microbial biodeteriogens live in close contact with the materials they damage, generally forming biofilms on the surface and often penetrating into the material. It is thus essential that any investigation into biodeterioration involve biofilm sampling. There is no standard method for this procedure. Some of the techniques used, with their advantages and disadvantages, are shown in Table 5.1.

Monitoring of a material or environment to assess the risk of biodeterioration will generally involve the analysis of samples taken over a period of time. In this case, it is important to use a well-standardized sampling technique so that results can be easily compared and an increase in the numbers of deteriogenic microorganisms quickly detected. The organisms that cause damage are normal environmental inhabitants, and low numbers might be expected to be found in "safe" situations; only if these numbers begin to increase must action be taken. Similarly, when the aim of sampling is to assess the efficacy of biocides, any method may be used as long as it is standard. Swabbing, for example, is almost impossible to standardize; the humidity of the swab and the pressure applied during sampling are difficult to keep constant.

More recently, methods which measure the numbers or activity of the relevant microorganisms directly in the biofilm have been developed. These real-time in-situ analyses may involve sophisticated techniques or may lack sensitivity, but, when optimized, offer an excellent tool for the biodeterioration scientist. One simple example is the use of an oxidoreduction reaction to measure microbial respiration. The colour change induced in the substrate has been used to estimate general microbial activity in biofilms on stone surfaces. Fluorescent dyes, such as acridine orange and 4',6-diamidino-2-phenylindole (DAPI), can be used to stain microbial cells in a thin biofilm, and these can then be counted directly by use of an epifluorescence microscope. Living cells can be counted by use of dyes which fluoresce only when acted on by microbial enzymes; 2'7' dichlorodihydrofluorescein (DCFH), for example, is supplied in the oxidized form and fluoresces green only when reduced by active cells. More sophisticated are modern methods of microscopy – the atomic force microscope, capable of resolving atoms on a surface, the environmental scanning

Table 5.1. Methods used for biofilm sampling

Method	Examples	Comments
Swabbing	Food-preparation surfaces	Simple, but difficult to standardize; only a fraction of the total organisms is collected. Some swabs are inhibitory.
Scraping	Pipelines ("pigging"), mineral materials	More efficient than swabbing, but extraneous materials (which may be inhibitory) are also collected.
Washing (agitation in liquid)	Fabrics, small, non-soluble specimens	A surfactant that is not antimicrobial should be included in the wash water.
Contact plate	Surfaces in buildings, paper	Allows the spatial arrangement within the biofilm to be maintained; suitable only for thin biofilms.
Adhesive tape	Glass, construction materials	As for contact plate, but more efficient at removing cells from the surface; tape must be non-inhibitory if cells are to be cultured.
Collection of substrate (material) with intact biofilm	Paint films, pipe sections, stored grain	The best of all possible worlds for the microbiologist, but may damage the material. Planning at the design stage of a system allows the incorporation of readily removable coupons.

electron microscope, which avoids the artefacts induced by sample preparation for normal SEM, and confocal laser SEM, which allows optical sections to be taken through the surface layers of a material and was indispensable in constructing our current 3-D model of the biofilm, with its stacks, pores, and channels. All of these are currently research, rather than routine, analytical tools. This is also the case with microsensors which can be used to measure pH, oxygen concentration, and so forth, in micrometre sections of a biofilm. Other examples are discussed under the Research section of this chapter.

In spite of the importance of biofilms, much biodeterioration testing still relies on detection of planktonic microorganisms. This is the traditional method of identifying an aggressive environment and of testing biocides for their efficacy. Sampling techniques for liquids and the statistics involved have been well described in the clinical literature and need not be discussed here.

Detection and enumeration

Standard microbiological methods of microscopy and culture are much used in biodeterioration testing. Any practical microbiology manual will contain the principles and basic techniques used. A more applied approach, which uses detection of microorganisms involved in biocorrosion as an example, can be found in Beech, Tapper, and Gaylarde (1998).

A number of kits, suitable for rough and rapid detection of the more economically important deteriogenic microorganisms, are available on the market. Fuel contaminants and SRB are good examples. Some of these are simple 'dip slides', plastic slides coated with appropriate growth media, which can be dipped into a liquid sample or pressed onto a solid surface before incubating to allow the development of colonies from the cells picked up during the sampling stage. Such methods require no expertise and no laboratory facilities, but generally have a rather low sensitivity. They can be made selective for a particular group of microorganisms by use of appropriate specific media.

Methods based on the detection of biomolecules, such as adenosine triphosphate (ATP), phospholipids, and ergosterol, can be used for rapid and quantitative detection of general or specific microbial contamination. ATP detection by measurement of bioluminescence has long been used in the food industry to assess microbial contamination of food preparation surfaces. The principle of the technique is shown in Figure 5.1. The greater

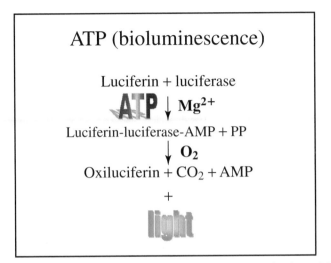

Figure 5.1. Principle of the bioluminescence technique for determining ATP.

the number of living (or recently living) cells, the greater the amount of ATP and the more light released for measurement in the luminometer.

Ergosterol measurement, as mentioned in the section on the environmental simulation test later in this chapter, is a useful rapid measure of fungal biomass. Ergosterol is a structural component of fungal membranes and is absent, or present in small amounts, in other organisms. Although the technique has its drawbacks [use of high-performance liquid chromatography (HPLC) and difficulty of relating ergosterol concentration directly with fungal biomass], it is, nevertheless, a useful guideline and certainly no less accurate than plating techniques. Determination of phospholipids, although a rapid method for biomass quantification, is not normally used routinely and is further discussed in the Research section of this chapter.

In 1979, Bailey and May examined the use of commercially available kits for detection of contamination in marine diesel and found that most gave results which correlate well with standard plate counts. In the ships' fuel tanks, <50–10^3 bacteria/ml fuel and 10^2–10^6 bacteria/ml seawater ballast, with corresponding fungal counts of 0–<10 and 20–10^2, were detected. A 'rapid' detection method used to examine jet fuel in Brazil is the Boron Microbe Monitor Test, in which fuel samples are incubated over an aqueous mineral salts layer with or without biocide. A positive result is one in which a change at the interface (fungal growth) is seen in the non-biocide-containing flask. Although simple to perform and interpret, the method

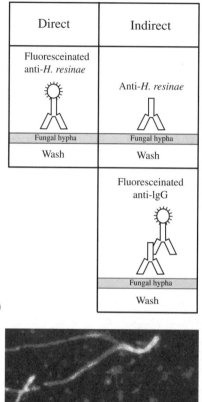

(a)

(b)

Figure 5.2. (a) Principle of the immunofluorescence technique, (b) *H. resinae* in a biofilm on metal, stained by the immunofluorescence method. Photo: Dr Christine Gaylarde.

gives final results only after 1–2 weeks, too long for an aircraft to remain out of action. A rapid immunofluorescence method designed to detect *Hormoconis resinae* specifically has been described (see Figure 5.2), but has not yet been adopted for routine use, probably because of the need for somewhat expensive apparatus (an epifluorescence microscope) and trained personnel. This highly sensitive technique gives results in an hour, a considerable improvement on standard methodologies.

An alternative immunological method, enzyme-linked immunosorbant assay (ELISA), is much used for rapid microbial detection and enumeration in the clinical area and has also been adapted for use with *H. resinae*. The principle of the technique is similar to that of the immunofluorescence method, but an enzyme is used as the marker molecule rather than a fluorescent compound. No epifluorescence microscope is required, a positive reaction being visualized as a colour change in the enzyme substrate, and the whole process is readily automated to give rapid quantitative results. Some immunological techniques have been developed into kits for the detection of deteriogenic microorganisms. One example is RapidCheck for SRB detection; this is an ELISA method measuring the SRB-specific enzyme, adenine phosphosulfate reductase.

Although immunological methods are highly sensitive and specific, they rely on biological means for the production of the most important reagents – the antibodies. The use of monoclonal antibody technology has facilitated the production of standard immunoglobulin reagents, but variation can still arise, and this limits, to some extent, the use of these methods. It seems likely that they will be replaced in the future with molecular biology techniques, involving nucleic acid analysis. This is an area of intense research at present, and an introduction to the topic is given in the Research section of this chapter.

BIODETERIORATION TEST TECHNIQUES

The testing of materials for their biological susceptibility falls into three main categories: the preliminary screen, the controlled environmental simulation, and the field trial or long-term exposure. Each category has advantages and disadvantages, and two or more categories are often combined or carried out sequentially. A fourth category should also be mentioned: the use of tests which comply with a national standard. These tests may be considered as belonging to either the first or the second category, but they have a special role in ensuring quality levels in materials and products. They are considered in a separate section.

The preliminary screen

In the early stages of the development of a product, there may be a number of promising formulations which merit further examination. For example, a new class of chemicals with predicted biocidal activity may have been synthesized or the efficacy of a range of preservatives in a new product may need to be carried out before the most appropriate is recommended for further evaluations. To remove unsuitable candidates which, in practice, do not demonstrate activity or perform adequately, a laboratory screening technique can be employed. Such screens are very attractive in that they can be carried out quickly with traditional microbiological techniques and a large number of replicates. Their limitations lie in the fact that they do not simulate the end use of the product and often they do not make allowances for ecological considerations. Screens are also open to erroneous interpretation and overstatement.

Such screens are often carried out in Petri dishes or shake flasks which contain media designed to create the optimum conditions for microbial growth. The preservative is either incorporated into the media after sterilization or, in the case of solid media, may be placed in a well cut into the agar. The test organisms are then inoculated into the media as either broth cultures or suspensions spread across the surface of the solid media. In liquid media a high log reduction (4 logs) may be the end point and in solid media a zone of inhibition around the dosed well. Alternatively, an aliquot of the product containing the biocide (Figure 5.3) is placed on the surface of the inoculated agar. The zone of inhibition of growth around the well or product is measured. The results are interpreted with the assumption that, the greater this zone, the greater the potency of the biocide. However, it may be argued that the greater the zone of inhibition, the greater the leach rate; an undesirable quality in a biocide required to have a long life. The constituents of the media may also affect the movement of the biocide, as will the carrier liquid.

The screens are often conducted at temperatures (25–30 °C) at which the growth rate is at a maximum. Laboratory strains of bacteria and fungi are used as pure cultures. These strains may be chosen because of their ubiquitous nature (e.g., *Pseudomonas*), the ease with which they are cultured in the laboratory (e.g., *Escherichia coli*), or their striking morphology in culture (the dark brown spore masses of *Aspergillus niger*). Sometimes, a range of microorganisms are chosen according to their taxonomic differences, or a 'pet' strain will be used which exhibits a particular property. In the latter case a cellulolytic fungus which grows well on laboratory media might be used to test the efficacy of a biocide to be used to protect cellulose or cellulose-derivative-containing products. Such a fungus may not be recognized as a colonizer of these products, and thus any extrapolation should be examined with care. Danger also lies in the validity of results from one microorganism in a screen. Copper tolerance, for example, is known to occur in *Aureobasidium pullulans* and *A. niger*. The sole use of either one of these fungi in a biocide screen involving a formulation containing copper might result in the rejection of a biocide suitable for application where these two fungi are unlikely to occur as biodeteriogens. The other possibility is that an unnecessarily high level of the biocide would be recommended for practical applications.

The ecology of the microbial community associated with the biodeterioration of a material cannot be considered in the Petri dish alone. It must be borne in mind that interactions may occur which enhance the breakdown rate in the material or adversely affect the potency of a biocide at a concentration known to be effective in pure culture. We often encounter examples of the latter possibility when proceeding from the initial screen to the use of mixed cultures or an inoculum isolated from an infected system (often referred to as 'house organisms'). The influence of the biocide on a biofilm attached to the internal circumference of a pipe cannot be assessed by normal methods. It is necessary to devise a technique to simulate the environment, or, at the very least, to place metal coupons, which can be removed for examination at set intervals, in the medium.

By far the most commonly used technique when the efficacy of preservatives in a particular material or product is assessed is to use the repeat challenge or insult test. The basic principle is that an inoculum, either mixed or single, of known titre is added to samples of the product, for example, a liquid paint, which contain various concentrations of the preservative under test. Together with a blank control (no preservative) and a reference system with a known end point (positive control), the inoculated samples are incubated at a set temperature (normally

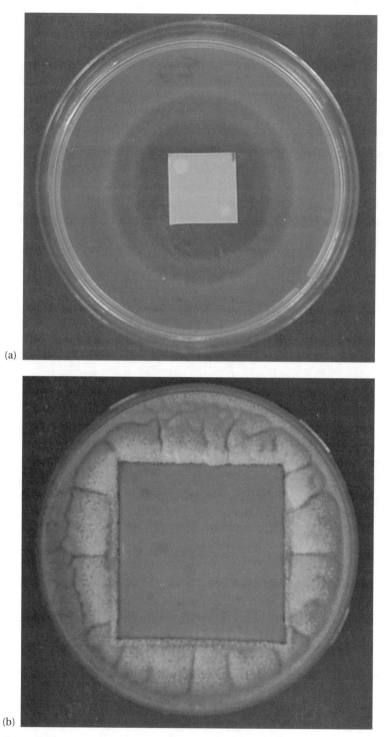

Figure 5.3. (a) Bacterial and (b) fungal zone of inhibition tests. Photos: Dr K. J. Seal.

25–30 °C) and aliquots removed after set periods of time (usually up to 7 days). These are spread across an agar plate, either straight from the sample containers or after a dilution series in Ringer's solution and the survivors either assessed using a subjective rating scheme or counted at a suitable dilution (normally between 30 and 300 colonies). After 7 days (this may be extended if the preservative shows a slower rate of kill), the samples are rechallenged or insulted again with the same inoculum and the process for detecting survivors repeated. The challenge may be repeated many times but the norm is three. A preservative able to kill three challenges is deemed to be worthy of further trials, normally at plant level (i.e., a works trial). There is much discussion, particularly among cosmetics microbiologists, as to the usefulness of a repeat challenge test over a single challenge if the purpose of the test is to determine efficacy. After all, three challenges of a 10^8 concentration of bacteria might be considered to be the same as one challenge of 3×10^8 bacteria if the capacity of the preservative is being assessed. However, the repeat challenge is accepted as a good predictor of performance in the field.

Now that the limitations of screening techniques have been listed, it should be noted that they do provide a quick indication of the gross effect occurring in the system (if designed to minimize or at least recognize the preceding points). Screening tests in biodeterioration usually use the growth/no-growth assessment. Some rating schemes may attempt to quantify growth by a subjective assessment as to whether growth is excellent, good, moderate, or poor. The amount of coverage of the sample or streak by the microorganism on a 0%–100%, a 0–5 scale, or an increasing number of pluses (− to +++) is often used as an indication of the efficacy of the preservative. This approach was recently criticized by Shirakawa et al. (2002) for fungal inocula, as degree of sporulation, an important indication of fungal reproductive capacity on the materials, is not considered.

Several test methods rely on the use of an agar medium containing mineral salts, to which the material is introduced as the sole carbon source. This may be an effective protocol for some materials, for example, textiles and leather, but some plastics, such as the polyurethanes, require the presence of supplementary nutrients such as sugars for initiating growth. In the absence of the supplements, no attack occurs for at least 28 days. The pH of the medium may affect the activity of the biocide being tested; for example, a reduction of the pH from 4.5 to 3.3 has been found to be sufficient to reduce the toxicity of the copper ion towards copper sensitive strains of *A. pullulans*. It is therefore important that the results are interpreted with care and restraint.

The environmental simulation test

These tests are designed to take into account the end use of the material while keeping a reasonable amount of control on the environmental conditions. To achieve these aims, the candidate biocide may be incorporated into the material which it is designed to protect, or a material may be exposed to the type of environment it is likely to encounter in service. The material, with or without biocide, may then undergo artificial weathering or ageing to simulate the effects of rain, temperature cycles, and sunlight. Weatherometers, for example, the xenon arc and quantum-UV types, are specially designed to carry out timed cycles (e.g., 100, 200, and 400 h) of the preceeding climatic conditions to produce samples for subsequent testing. It is also possible to simulate an industrial atmosphere by the introduction of sulfur dioxide and ozone. The sample may be exposed to timed cycles of temperature and relative humidity (RH) to simulate a tropical environment. Such conditions rarely appear in published methods, but are worth serious consideration. Cement-based products which may require direct evaluation, or the testing of a surface coating, can be carbonated before testing to reduce the alkaline conditions (pH 10–11) of the fresh product, which would lead to erroneous results. The materials will almost certainly encounter some form of contamination, which may act as a nutrient for any potential colonizer. To account for this factor, the material may be sprayed or dipped in a suitable mineral nutrient solution to encourage growth. When this is employed, a visual assessment alone is not sufficient unless the material contains a candidate biocide and inhibition of growth is used as the assessment criterion. Some form of physical test to examine adverse changes in the properties of the material must be used. This aspect will be discussed later.

Two environments are used as representative of the wide variety of end uses for materials. These are the soil and a humid atmosphere. Both of these conditions can be set up in the laboratory with the minimum of equipment. The soil can be obtained from the garden, or in the form of a pre-mixed compost such as John Innes no. 1 (note that the 'sterilized' label on such soils refers to weed seeds and insects and not to microorganisms). The soil should have an active microbial flora, a pH of between 5.5 and 7, be of a loam composition, and have a moisture content of 25%–35%,

Figure 5.4. Materials undergoing a soil burial test. Note the piece of degraded cotton textile which has been used to evaluate the activity of the soil. Photo: Dr K. J. Seal.

or 60% of its maximum water-holding capacity. Sieving through a garden sieve is recommended to remove stones. The soil is usually placed in trays 15–24 cm deep and maintained at a constant temperature. The trays of soil are often termed soil burial beds (Figure 5.4).

They can be quite large, occupying a room specially constructed for the purpose, with a controlled environment. The beds may be kept for several years if they are in regular use. The occasional addition of organic matter or fertilizer may be necessary if activity begins to wane. For shorter-term use, large beakers, buckets, or sandwich boxes can be employed, but their limitations in representing only a diverse source of microorganisms and not a complete natural soil environment should be borne in mind. As the name of the beds implies, materials for test are buried in the soil for predefined periods of time. Three months is often quoted, but it is important to allow for retrieval at other times by the burial of extra samples which can be retrieved over shorter and longer periods of exposure. Soil is recognized as a severe or aggressive environment, and its use is justified for materials and preservative systems which will come into contact with the soil during their service life. Microbiologically, it is an ill-defined system

Figure 5.5. Vermiculite bed testing of dry-film paint algal preservatives. Photo: Dr K. J. Seal.

but, surprisingly, if run well, soil burial techniques yield comparable results and are employed as part of a number of National Standards in Europe and the United States.

Many materials will be placed in that part of the environment distant from direct contact with the soil, yet subject to colonization by soil and air-borne microflora. In these situations, it is the humidity of the atmosphere which plays an important part in the germination and sustained growth of the potential biodeteriogens. There are many industrial and domestic situations in which localized high-humidity conditions can occur. In these situations, algae or fungi are the predominant colonizers and defacement is the major effect. This can be simulated in the laboratory by use of chambers containing either water, to give a saturated atmosphere of 100% RH, or a range of saturated salt solutions for controlled humidities of less than saturation. The materials under test are then suspended in the atmosphere or placed on wetted vermiculite (see Figure 5.5), where they are sprayed with a mixed inoculum of fungi and/or algae and incubated at a constant temperature. After a predetermined time (often 28 days), the materials are examined for the presence of active growth on the surface. The extent of the growth is assessed in terms of either percentage cover or density, as described previously. A recent report suggests ergosterol analysis by HPLC to give a quantitative measure of fungal growth on paint films. Extremely rapid and simple, the method showed the relative efficacy of fungicides incorporated into the paint film.

Modifications of this system may be used to meet particular situations. One such system is used to test paint film fungicides. A painted surface may be subject to condensation as a result of a temperature gradient through a wall and this can encourage mould growth (typical in bathrooms and poorly ventilated kitchens). To simulate this, the coated panels are suspended in a chamber with water in the bottom heated to 5 °C above the ambient temperature. This induces water condensation on the film. A mixed inoculum is applied to the coating and assessed for up to 3-month incubation. Damp vermiculite beds are now used routinely as a technique to simulate a surface which is continuously exposed to damp conditions. This technique is particularly suited to the testing of film algicides. Flat panels (mineral-based board or softwood), often previously weathered as already described, with the paint film under test applied to the upper surface, are laid flat on the wet vermiculite in a clear plastic chamber (sandwich boxes with close-fitting lids are perfect). The mixed algal inoculum is applied (brushed or sprayed) and assessments made according to the amount of growth present for up to 3 months. Timed cycles of light and dark are used to simulate the diurnal cycle, and periodic additions of mineral nutrients may be sprayed onto the inoculated panels.

When a particular type of biodeterioration is clearly defined and the causative organisms have been identified, for example, wood decay, an acceptable simulation is to use those organisms in conjunction with susceptible timber species in order to test the effectiveness of wood preservatives by measuring loss in weight of impregnated wood blocks (compared with controls) after 3-month exposure to pure cultures of wood-destroying fungi. Conditioning of the blocks by use of a leaching step eliminates those biocides which are likely to have a short residence time and hence limited performance life as preservatives. The value of the simulation test lies mainly in its more realistic prediction of how a material will perform in service while the environment is kept under close control. This enables the results of tests carried out in different laboratories and at different times to be compared.

The field trial

The field trial is usually a long-term test (greater than 1 year), in which the material is exposed to one or more service situations or a range of climates in which the material is likely to be placed when in service. Field trials may last for up to 10 years and encompass climates ranging from humid tropical conditions to freeze/thaw situations. The design of a field trial is of paramount importance in deciding the exposure sites, the best way to present the material to the environment, the number of samples, the frequency of sample examination, the properties to be tested, which organisms it will be necessary to monitor, whether photographic data might be useful, what background data are necessary (meteorological records of the site, surface temperatures), who will administer and coordinate and ensure continuity over the whole trial, and whether there should be a master record or manual of procedures which describes the whole field trial to ensure traceability and where it should be placed for safekeeping. In selecting sites, it is important that their security is established to prevent loss of samples. Contingency plans should be considered in the event that a site is closed down or transferred. Arrangements for the transport of samples from the site to the point of examination must also be arranged, if this cannot be completed by the investigator. The organization and setting up of a field trial can thus be a very time-consuming and costly business, especially where a material is to be used in a variety of situations for which an interaction between physical and biological factors is likely. It is not possible to generalize on the experimental design of a field trial. However, a brief description of the type of trial often used to test the effects of the environment on materials used either as external building materials or surface coatings will serve as an example.

The surface-coatings industry and those industries which produce mineral-based composite boards used for external cladding and roofing use exposure racks on which test panels can be mounted. The way in which panels are mounted is important with respect to their angle to the horizontal plane, the direction in which the panel faces, and the distance from the ground. Panels are usually inclined at 45° or 90° to the horizontal plane, and in a north- or south-facing aspect (Figure 5.6).

A panel at 45° will be the most severely exposed, as contamination falling on it will remain on the surface for a longer period than one which is vertical and will, if it is also south facing in the Northern Hemisphere, catch more of the sun. Elaborate methods for rotating panels continuously to keep them facing the sun have been devised. However, UV measurements show that reflected and incidental light concentrations on static panels at 45° make rotational methods an unnecessary luxury. The north-facing panel, in the Northern Hemisphere, does not receive direct sunlight and is likely to remain cooler and damper than the south-facing sample. This

Figure 5.6. Painted panels exposed at 45° on a south-facing aspect, being examined by one of the authors (DA). Photo: Dr K. J. Seal.

often results in algal and fungal growth, which can affect some of the properties of the samples in time. These orientations are reversed in the Southern Hemisphere. The height at which the panels are placed above ground level can affect the source of contamination. To simulate a non-soil-contact situation the panels need to be at least 1 m above ground level to prevent splashes. Panels may, of course, be placed at heights less than this in order to allow for soil contamination. The panels may also be sheltered by a roof over the whole rack, if this is required for simulating a less exposed application.

The rack design is very important in that it must be able to support securely the total number of panels to be exposed, it should not decay or corrode significantly over the period of the trial, its design should not contribute to the changes observed in the panels unless this has been deliberately incorporated, and it should be designed so that it is easily maintained. It should be made from preserved timber or rust-proofed metal and have supports which enable 45° and vertical fixings in two opposing directions. Panels are attached to the rack by fixings which do not affect the material by, for example, enhancing the corrosion of a metal panel being tested; in this case fixings should be coated to prevent corrosion. The periodic removal of panels is followed by observation of the surfaces and identification of microorganisms present; relevant mechanical or chemical tests are required for determining any changes in the exposed material.

Statistical analysis is important in the interpretation of these tests. Sufficient replicates should be prepared to ensure that this is possible, and non-parametric statistical analysis is likely to be the most appropriate. Expert advice should be taken in the absence of a standard method. A recent paper on biodeterioration of external paint films with and without biocides suggests the use of logistic regression analysis with the use of accumulated data to avoid the necessity for large numbers of replicates over a time period. A knowledge of statistics is extremely important when such comparative studies are being planned to ensure that the results will be meaningful.

Standard tests in biodeterioration

There are a small number of National and International Norms (Standards) which contain procedures to evaluate the effects of organisms on materials or the preservative systems used in those materials. A list of those standards in current use can be found in Tables 5.2 and 5.3, from which it can be seen that they fall into either the screening or the environmental simulation category.

The screening of materials for their biological resistance or susceptibility involves placing the sample on a nutrient-containing agar, in which the material is often the 'sole' carbon source, and spraying it with a single or mixed culture of either bacteria or fungi. The strains used are specified by a culture collection catalogue number, although the standard often allows other strains to be used if there is agreement between the parties involved. The susceptibility is judged by the amount of growth on the sample after a suitable incubation period or, in the case of preservatives, by absence of growth on the sample, or an inhibition zone around the sample. Weight losses in preweighed and conditioned wood blocks or plastic samples are also used as an indication of the efficacy of a preservative or the susceptibility of the test material to microbial utilization. The information gained from this type of standard relates solely to the biodeterioration potential of raw materials and is not an in-service simulation test in which complete products or components comprising a number of raw materials are tested. In the latter case, the components are either hung in a high-humidity chamber or buried in soil, as previously described, and an assessment is

Table 5.2. British and European biodeterioration testing standards

Number: Year	Title
BS1982:1990	Methods of test for fungal resistance of panel products made of or containing materials of organic origin
BS2011:1989	The environmental testing of electronic components and electronic equipment. Part 2.1, Test J. Mould growth
BS2087:1992	Preservative textile treatments
BS3046:1981	Specification for adhesives for hanging flexible wall coverings. Appendix G. Test for susceptibility to mould growth
BS3900:1989	Methods of test for paints. Test G6. Assessment of resistance to fungal growth
BS4249:1989	Specification for paper and cork jointing. Section 5.7. Resistance to mould growth
BS5763:1996	Methods for the microbiological examination of food and animal feedingstuffs
BS5980:1980	Specification for adhesives for use with ceramic tiles and mosaics. Section 7. Resistance to mould growth
BS6068:1988	Water quality. Guide to the enumeration of microorganisms by culture
BS6085:1992	Methods for the determination of the resistance of textiles to microbiological deterioration
BS6920:2000	Suitability of non-metallic products for use in contact with water intended for human consumption with regard to their effect on the quality of water. Part 2.4. Growth of aquatic microorganisms
BS7066:1990	Wood preservatives. Effectiveness against blue stain
BS7874:1998	Method of test for microbiological deterioration of elastomeric seals for joints in pipework and pipelines
BSEN 113:1997	Test method for determining the protective effectiveness against wood-destroying basidiomycetes
BSEN 152:1988	Laboratory method for determining the protective effectiveness of a preservative treatment against blue stain in service. Part 1. Brushing procedure. Part 2. Methods other than brushing
BSEN1040:1997	Chemical disinfectants. Basic bactericidal activity. (See also BSEN 1275, 1276, and 1650)
BSENISO846:1997	Plastics: Evaluation of the actions of microorganisms
DDEN839:2002	Wood preservatives. Effectiveness against wood-destroying basidiomycetes
BSISO14851:1999	Determination of ultimate biodegradability of plastic materials in an aqueous medium. (See also BSISO 14852 and 14855)

BS, British Standard; EN, European Norm; DD, Draft Development; ISO, International Standards Organisation.

Table 5.3. USA biodeterioration testing standards

Number	Title
ASTM D2574–00	Standard test method for resistance of emulsion paints in the container to attack by microorganisms
ASTM D3273–00	Standard test method for resistance to growth of mould on the surface of interior coatings in an environmental chamber
ASTM D3274–95	Standard test method for evaluating the degree of surface disfigurement of paint films by microbial (fungal or algal) growth or soil and dirt accumulation
ASTM D4300–01	Standard test methods for the ability of adhesive films to support or resist the growth of fungi
ASTM G21–96	Standard practice for determining resistance of synthetic polymeric materials to fungi
ASTM G29–96	Standard practice for determining algal resistance of synthetic polymeric materials
MIL STD 810E	Environmental test methods. Method 508.2. Fungus

ASTM, American Society for Testing and Materials; MIL STD, Military Standard.

made of the extent of growth over the surface. A performance test is often carried out on completion of the microbiological test.

It is thus important in drawing up test specifications which include biological testing that the correct standard to identify the extent of biodeterioration is employed. For example, if the components of a building board containing cellulose pulp, mica, and cement were tested, the cellulose component would invariably fail, but as a composite it might pass a soil burial or a high-humidity test because of the alkaline pH of the system. Standards which test the biodeterioration of materials are in need of much improvement and update. Some can give erroneous results, whereas others employ a range of organisms which are difficult to justify as direct biodeteriogens for the materials they are designed to evaluate. The requirement of the Biocidal Products Directive (see Chapter 6) to demonstrate efficacy in preservatives will mean the introduction of more focused test methods.

National laboratory accreditation schemes, set up to ensure standards in quality of testing by laboratories, have now been implemented in many countries. The UK has introduced the United Kingdom Accreditation

Service (UKAS), which encompasses both testing and calibration laboratories. Of these, only a very small number currently carry out biodeterioration testing; by far the majority of testing is carried out as part of technical
support services supplied by the preservative/biocide companies.

RECENT RESEARCH TECHNIQUES IN BIODETERIORATION

The standard tests and treatment methods described in the preceding
section are subject to constant revision and updating. Research on the
theoretical and applied aspects of biodeterioration is fundamental to this
process. Research directed specifically to microbial deterioration of materials is concerned mainly with the elucidation of mechanisms and the
identification and physiology of the organisms involved, the eventual aim
being control or prevention of damage. Particularly active areas recently
have been related to biofilms and to the development of molecular techniques for the study of microbial deteriogens.

Nucleic acid analysis techniques

The development of the polymerase chain reaction (PCR) (see Figure 5.7) in
the early 1980s led to a revolution in our approach to biological research.
Primers 1 and 2 in Figure 5.7 are synthetic oligonucleotides which bind
specifically to selected sites on the DNA, allowing the Taq enzyme to begin
replication. Because replication is always in the same direction on the DNA
strand (5' to 3'), a separate primer is required for each individual strand.
These are known as the forward (F) and reverse (R) primers.

   The ability to amplify parts of the genome specifically means that gene
probes can be readily produced to detect microbial genera or species and
their activities. Manipulation of the primer sequence and/or the conditions of the annealing step in the PCR allows us to change its speci-
ficity, decreasing it to allow concurrent detection of a general group of
organisms (e.g., fungi in general), or increasing it to limit the types of
organisms/activities recognized. Lipase genes can be identified to indicate
organisms capable of breaking down fats, cellulase genes for cellulolytic
cells, and so on. Targeting mRNA, rather than DNA, in the cells allows us to
determine whether the genes are active. This is the RT (reverse transcrip-
tion) PCR.

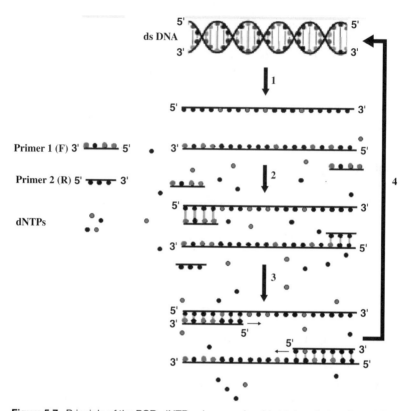

Figure 5.7. Principle of the PCR. dNTPs, deoxynucleoside triphosphates. Step 1. Denaturation of the double-stranded (ds) DNA by raising the temperature to 90–95 °C. Step 2. Annealing of the primers to the separated DNA strands. Temperature may vary between 40 and 70 °C. Step 3. Extension of the newly forming DNA chains by the enzyme Taq polymerase (temperature usually 72 °C). Step 4. Repeated cycles (up to 35) to produce multiple copies of the ds DNA PCR product.

The PCR, along with methods of analysing the amplified DNA, such as amplified 16S rDNA restriction analysis (ARDRA), single-strand conformational polymorphism (SSCP), and denaturing gradient gel electrophoresis (DGGE), has already shown the presence of previously unknown microorganisms in deteriorated materials. It is now regarded as axiomatic that only 1%–10% of all microorganisms have been cultured in artificial media. When DNA sequences from the gene for the small subunit (ssu) of ribosomal rRNA are used as primers (16S rDNA in prokaryotes, 17S or 18S rDNA in eukaryotes), microorganisms can be detected and

Table 5.4. Examples of specific primers used in the PCR

	Specificity	Base sequence
ssu rDNA Primers		
ITS1F	Fungi	CTTGGTCATTTAGAGGAAGTAA
ITS4 (R)	Universal fungal	TCCTCCGCTTATTGATATGC
ITS4-A (R)	Ascomycetes	CGCCGTTACTGGGGCAATCCCTG
ITS4-B (R)	Basidiomycetes	CAGGAGACTTGTACACGGTCCAG
27F1	Universal bacterial	AGAGTTTGATCCTGGCTCAG
408R	Cyanobacteria	TTACAA(CT)CCAA(AG) (AG)(AG)(AG)CCTTCCTCCC
1492R	Universal bacterial	GGTTACCTTGTTACGACTT
63F	Spirochaetes	CATGTCGACGTYTTAAGC ATGCAAGT
NSM156 (F)	*Nitrosomonas* spp.	TATTAGCACATCTTTCGAT
Examples of primer pairs for genes involved in biodeterioration		
opl1	Lipases	GAATTTTTTTAGGAGGACACT ATGAGTC (F) CCCTGCAGCTTATTTACCC CCATTATAAGTGGTTTGATTTTAGGCC (R)
catA	Catechol-1,2-dioxygenase (hydrocarbon degradation)	GAAGGACCGCTATATGTTGC AGGTGC (F) TAGTGAATATGCGCAGGG CG (R)

F, forward primer; R, reverse primer.

identified rapidly and specifically without the need for culture. Once a specific DNA region (probe) has been identified for a "new" organism, fluorescent insitu hybridization (FISH) may be utilized to visualize the cells in their natural habitat. The technology has been well developed for bacteria, but research on other microorganisms is still lacking. Table 5.4 shows some of the primers which have been used to amplify various types of DNA.

Identification at genus or species level can ideally be achieved by the determination of the nucleic acid base sequence of the amplified DNA. This is then submitted to the available databanks (e.g., www.ncbi.nlm.nih.gov/BLAST) for comparison with known sequences which have been deposited by previous workers. One of the problems is that these databases are limited, especially in the case of fungi and cyanobacteria, and many submitted sequences will find no good match and remain unidentified. One of the most important current research thrusts must be the sequencing of the DNA of more microorganisms, both cultured and uncultured. Whole genome sequencing is important, but even the deposition of partial sequences, determined by the use of only a small number of primers, will lead to an improvement in the services offered by the databanks. A constantly updated list of completed sequences of microbial genomes can be found at http://www.tigr.org/tdb/mdb/mdbcomplete.html.

Community analysis

Biodeterioration is all about interactions between living organisms and inert materials in the natural environment. Because organisms are not normally present in pure cultures in nature, it is important to examine the whole living community and its relationship with deterioration. This has been done in the past by use of techniques such as phospholipid fatty acid analysis. The determination of specific fatty acids by gas chromatography and mass spectroscopy (GC-MS) can indicate not only the amounts, but also the types, of microorganisms present. So-called 'signature fatty acids', some of which are shown in Table 5.5, allow the quantitative detection of specific groups of microorganisms in an environmental sample. However, rather lengthy and specialized extraction and separation techniques are required, and marker phospholipid fatty acids (PLFAs) have not been identified for all groups of deteriogenic organisms; thus the new molecular biology techniques make concurrent detection of microorganisms in a community a much more feasible proposition.

Microbial populations in any environmental niche can be studied more readily, rapidly and accurately than in the old 'pre-PCR' days by use of community analysis methods such as DGGE and temperature gradient gel electrophoresis (TGGE) of amplified DNA. These two techniques allow the separation on a polyacrylamide gel of DNA fragments with different base sequences. The temperature or concentration of denaturing reagents (formamide, urea) at which the DNA strands separate depends on the base sequence. The two strands are prevented from separating completely (and being lost from the gel) by a 'GC-clamp' fixed to one end. This string of guanine and cytosine bases holds the two DNA strands together even at high temperatures or denaturing concentrations. Thus total DNA from a mixed population can be amplified with a rather general primer pair (for eubacteria, for example) and the amplified DNA from different bacteria, which are all similar sized fragments, indistinguishable on a normal agarose gel, can

Table 5.5. Some signature phospholipid fatty acids (PLFAs) used to detect specific groups of microorganisms

Signature lipid(s)	Microbial group
18:2ω6, 18:3	Eukaryotes
18:2ω6,9	Fungi
20:5ω3, 18:3ω3	Algae
10Me18:0	Actinomycetes
Terminally branched saturated PLFAs (e.g., i15:0, i17:0)	Anaerobic gram-negative bacteria
10Me16:0, i17:1ω7	SRB
i17:1ω7c	Desulfovibrio

then be separated on the DGGE/TGGE gel (Figure 5.8). Separate bands can be excised and cloned or further amplified for sequencing, allowing rapid determination of the diversity (and, it is hoped, identities) of microorganisms in the sample. Phylogenetic analysis based on the experimentally determined sequences together with those previously deposited can indicate the degree of relatedness between the microorganisms present and those already known.

This approach has been used to show that biodiversity in Arctic soils increased in response to petroleum contamination and that the majority of the community members belonged to the high G + C gram-positive bacteria. The range of activities in the community can be investigated by isolation and amplification of the mRNA (or, more easily, total RNA) present. In this case, the use of DNA microarrays (chips) is likely to be of great use. Microarrays are ordered arrangements of multiple DNA probes fixed to a small surface (the microchip). Over 90,000 probes can be affixed to a single chip. Figure 5.9 shows schematically how such an array might be designed and used. They are being applied to characterize complex community structure and function. For example, a basic array of catabolic genes derived from microbial genes for degradation of organic pollutants and genes involved in geochemical cycles is being used to monitor biodegradation potential in situ in contaminated environments. Protein chips have also been produced, but, as yet, not widely used. The reader is referred to the excellent *Scientific American* article on DNA chips (Friend and Stoughton, 2002) and invited to consider how this technology might be applied in the field of biodeterioration.

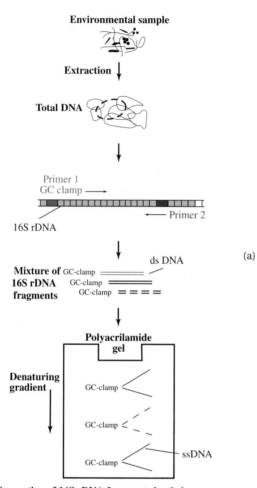

Separation of 16S rDNA fragments by their sequences

Figure 5.8. Gradient gel electrophoresis technique for separating DNA fragments with differing base sequences: (a) diagram showing the steps involved (modified from Amann and Ludwing, 2000),

Peptide nucleic acid probes

These DNA mimics were developed in the 1990s and promise to be even more useful than DNA probes, although they do not function as primers and cannot be used in the PCR. However, peptide nucleic acid (PNA) assays go beyond the possibilities of DNA probes because of their improved hybridization characteristics and capability of penetrating cell membranes. They are particularly useful in FISH and microarray technologies.

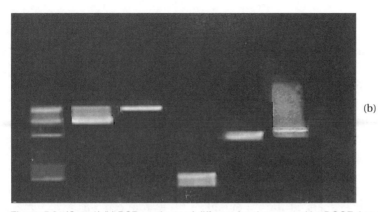

Figure 5.8. (*Contd.*) (b) PCR products of different fungi separated by DGGE. Lane number: 1, mixed fungi; 2, *Penicillium*; 3, *Cladosporium*; 4, *Alternaria*; 5, *Aspergillus*; 6, *Acremonium*. Photo: Dr Denise Saad.

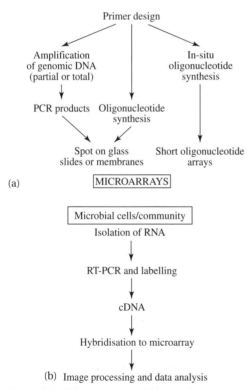

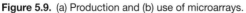

Figure 5.9. (a) Production and (b) use of microarrays.

Proteomics

The comparison of protein, rather than DNA, sequences is an alternative approach to the identification of relevant activities in deteriogenic organisms. Proteomics involves identifying proteins, outlining their three-dimentional structure, and determining how they interact. The proteome is much more complicated than the genome and offers greater possibilities to the investigating scientist, along with much greater challenges! A single gene can produce multiple gene products, enabling more information to be gathered about a particular system.

Once the genomic DNA of an organism has been completely, or substantially, sequenced, the proteins likely to be produced can be calculated, based on our knowledge of the working of the bacterial genome. These protein sequences can then be compared with a database of enzyme sequences to determine whether the bacterium under investigation has the ability to produce a protein similar in structure to a known enzyme. Only the sequences forming the active site of the enzyme need be known for this procedure to produce an exciting result, indicating that an organism isolated from a particular sample has the genetic information required for breaking down that substrate. COGs (clusters of orthologous groups of proteins) encoded in the genomes of various microorganisms represent functional pathways or systems, such as DNA repair and ion transport, and can be found at http://www.ncbi.nlm.nih.gov/COG.

Surface science techniques

A number of extremely sensitive instrumental techniques for the analysis of surfaces is now being used in biodeterioration. These include the microscopes mentioned in the previous section on detection techniques, and are not further discussed. The use of these, or of normal optical microscopes, in combination with instruments enabling the chemical analysis of surfaces on a microscale, allows us to correlate areas of damaged material with the presence of cells or their metabolic products, hence relating microbial activities directly with deterioration. Techniques that have been used in this way include energy-dispersive x-ray analysis, infrared spectroscopy, Fourier transform infrared spectroscopy, x-ray diffraction, x-ray photospectroscopy, x-ray fluorescence, surface-enhanced infrared absorption reflectance microspectroscopy, and time-of-flight secondary ionization mass spectroscopy.

Microsensors

The development of probes of a few micrometres in diameter has enabled measurements of pH and various ions to be made within the biofilm. Obviously such technology will be of immense importance in relating microbial activities to materials deterioration and its prevention. They have already been used, for example, to determine the concentration of chlorine in an artificial biofilm containing *Pseudomonas aeruginosa* and *Klebsiella pneumoniae*. Chlorine concentrations measured in the biofilm were typically only 20% or less of the concentration in the bulk liquid. Oxygen, pH, hydrogen sulfide, ammonium ion, nitrate ion and nitrate ion microsensors have been used to examine the dynamics of nitrification and of photosynthesis in biofilms. However, even these small sensors are considerably larger than a typical bacterial cell and their insertion into the biofilm must cause some disruption; hence the technique is invasive and artefacts may be produced. This must be considered when results are interpreted in relation to biodeterioration of the underlying surface.

THE FUTURE

Many of the technologies described in this chapter are already in routine research use in biodeterioration. Others still require considerable development before they are applied. The DNA- and RNA-based PCR techniques are readily developed into kits for identification of species and activities; microarrays are likely to be increasingly used here. The science of proteomics is still in its infancy, but growing fast. It offers exciting opportunities for the study of biodeterioration mechanisms and will doubtless be employed in the future. However, there will always be a place in research and routine investigations for the classical methods of microbiology.

WEBSITES

http://www.mcs.net/~ars/analytic/sem.htm (scanning electron microscopy) http://www.ambion.com/explorations/rtpcr.html (animated RT-PCR)

REFERENCES AND FURTHER READING

Amann, R. and Ludwig, W. (2000). Ribosomal RNA-targeted nucleic acid probes for studies in microbial ecology. *FEMS Microbiol. Rev.*, **24**, 555–65.

Arnason, T. S. and Keil, R. G. (2001). Organic–mineral interactions in marine sediments studied using density fractionation and x-ray photoelectron spectroscopy.
Org. Geochem., **32**, 1401–15.

Bailey, C. A. and May, M. E. (1979). Evaluation of microbiological test kits for hydrocarbon fuel systems. *Appl. Environ. Microbiol.*, **37**, 871–7.

Beech, I. B. (1998). Modern techniques for studying biofilm-induced corrosion.
NACE Latin American Region Corrosion Congress 6, Cancun, Mexico, Houston,
NACE International, 1998. Published on CD-ROM.

Beech, I. B. and Gaylarde, C. C. (1999). Recent advances in the study of biocorrosion –
an overview. *Rev. Microbiol.*, **30**, 177–90.

Beech, I. B., Tapper, R. and Gaylarde, C. (1998). Microbiological methods for the
study of biocorrosion. In *Biodeterioration des Materiaux*, Lemaitre, C., Pebere,
N., and Festy, D., (Eds.). EDP Sciences, Paris.

Bond, P. L. and Banfield, J. F. (2001). Design and performance of rRNA-targeted
oligonucleotide probes for in situ determination and phylogenetic identification of microorganisms inhabiting acid mine drainage environments. *Microbial.
Ecol.*, **41**, 149–61.

Bridge, P. D., Arora, D. K., Reddy, C. A., and Elander, R. P. (1998). *Applications of PCR
in Mycology*, CAB International, Wallingford, UK.

Davies, D. G., Parsek, M. R., Pearson, J. P., Iglewski, B. M., Costerton, J. W., and
Greenberg, E. P. (1998). The involvement of cell-to-cell signals in the development
of a bacterial biofilm. *Science*, **280**, 295–8.

De Beer, D., Srinivasan, R., and Stewart, P. S. (1994). Direct measurement of chlorine
penetration into biofilms during disinfection. *Appl. Environ. Microbiol.*, **60**, 4339–
44.

Ezzell, C. (2002). Proteins rule. *Sci. Am.*, **286**, 27–33.

Felske, A., Akkermans, D. L., and de Vos, W. M. (1998). Quantification of 16S rRNA in
complex bacterial communities by multiple competitive reverse transcription-
PCR in temperature gradient gel electrophoresis fingerprints. *Appl. Environ. Microbiol.*, **64**, 4581–7.

Friend, S. H. and Stoughton, R. B. (2002). The magic of microarrays. *Sci. Am.*, **286**,
34–41.

Galperin, M. Y. and Koonon, E. V. (1999). Functional genomics and enzyme evolution. Homologous and analogous enzymes encoded in microbial genomes.
Genetica, **106**, 159–70.

Garrity, G. M. (Series Ed.). (2001). *Bergey's Manual of Systematic Bacteriology*, 2nd
ed. Springer-Verlag, New York, Vol. 1, *The Archaea and the Deeply Branching and
Phototrophic Bacteria*, Boone, D.R. and Castenholz, R.W. (Eds.).

Gaylarde, C. and Cook, P. (1990). New rapid methods for the identification of
sulphate-reducing bacteria. *Int. Biodeterior.*, **26**, 337–45.

Gaylarde, P. M., Shirakawa, M. A., John, V., Gambale, W. and Gaylarde, C. C. (2003)
Statistical analysis of fungicide activity in paint films on two buildings. Submitted
to *Surface Coatings International*.

Graves, P. R. and Haystead, T. A. J. (2002). Molecular biologist's guide to proteomics.
Microbiol. Mol. Ecol. Rev., **66**, 39–63.

Guezennec, J., Ortega-Morales, O., Raguenes, G., and Geesey, G. (1998). Bacterial colonisation of artificial substrates in the vicinity of deep-sea hydrothermal vents. *FEMS Microbiol. Ecol.*, **26**, 89–99.

Heuer, H. and Smalla, K. (1997). Application of denaturing gradient gel electrophoresis and temperature gradient gel electrophoresis for studying soil microbial communities. In *Modern Soil Microbiology*, van Elsas, T. D., Wellington, J. D., and Trevor, E. M. H. J. (Eds.). Marcell Dekker, New York, pp. 353–73

Holman, H.-Y. N., Perry, D. L., and Hunter-Cevera, J. C. (1998). Surface-enhanced infrared absorption-reflectance (SEIRA) microspectroscopy for bacteria localization on geologic material surfaces. *J. Microbiol. Methods*, **34**, 59–71.

Innis, M. A., Gelfand, D. H., Sninsky, J. J., and White, T. J. (1990). *PCR Protocols. A Guide to Methods and Applications.* Academic Press, San Diego, CA, pp. 315–22.

Kim, S. H., Uzunovic, A., and Breuil, C. (1999). Rapid detection of *Ophiostoma piceae* and *O. quercus* in stained wood by PCR. *Appl. Environ. Microbiol.*, **65**, 287–90.

Kocher, T. D. and Wilson, A. C. (1991). DNA amplification by the polymerase chain reaction. In *Essential Molecular Biology: A Practical Approach*, Brown, T. A. (Ed.). IRL Press, New York, Vol. II.

Lawrence, J. R., Korber, D. R., Hoyle, B. D., and Costerton, J. W. (1991). Optical sectioning of microbial biofilms. *J. Bacteriol.*, **173**, 6558–67.

Lhoest, J. B., Wagner, M. S., Tidwell, C. D., and Castner, D. G. (2001). Characterization of adsorbed protein films by time of flight secondary ion mass spectrometry, *J. Biomed. Mater. Res.*, **57**, 432–40.

Lopes, P. T. C. and Gaylarde, C. C. (1996) Use of immunofluorescence to detect *Hormoconis resinae* in aviation kerosine. *Int. Biodeterior. Biodeg.*, **37**, 37–40.

Marina, M. A. and Blanco López, M. C. (2001). Determination of phosphorus in raw materials for ceramics: Comparison between x-ray fluorescence spectrometry and inductively coupled plasma-atomic emission spectrometry. *Anal. Chim. Acta.*, **432**, 157–63.

Muyzer, G., Waal, E. C., and Uitterlinden, A. G. (1993). Profiling of complex microbial populations by denaturing gradient gel electrophoresis analysis of polymerase chain reaction-amplified genes coding for 16S rRNA. *Appl. Environ. Microbiol.*, **59**, 695–700.

Palla, F., Federico, C., Russo, R., and Anello, L. (2002). Identification of *Nocardia restricta* in biodegraded sandstone monuments by PCR and nested-PCR DNA amplification. *FEMS Microbiol. Ecol.*, **39**, 85–9.

Pringault, O., Epping, E., Guyoneaud, R., Khalili, A., and Kuhl, M. (1999). Dynamics of anoxygenic photosynthesis in an experimental green sulphur bacteria biofilm. *Environ. Microbiol.*, **1**, 295–305.

Rautureau, M., Cooke, R. U., and Boyde, A. (1993). The application of confocal microscopy to the study of stone weathering. *Earth Surf. Processes Landforms*, **18**, 769–73.

Röllecke, S., Muyzer, G., Wawer, C., Wawer, G., and Lubitz, W. (1996). Identification of bacteria in a biodegraded wall painting by denaturing gradient gel eletrophoresis of PCR-amplified gene fragments coding for 16S rRNA. *Appl. Environ. Microbiol.*, **62**, 2059–65.

Saad, D. S., Kinsey, G., Paterson, R., Kim, S., and Gaylarde, C. (2001). Molecular methods for the analysis of fungal growth on painted surfaces. In *Proceedings 4LABS, the 4th Latin American Biodeterioration and Biodegradation Symposium*, 16–20 April 2001, Buenos Aires. Published on CD.

Sais-Jiminez, C. (Ed.). (2003). *Molecular Biology and Cultural Heritage*. Swets and Zeitlinger, Lisse, The Netherlands.

Schmidt, O. and Moreth, U. (2000). Species-specific PCR primers in the rDNA-ITS region as a diagnostic tool for *Serpula lacrimans*. *Mycol. Res.*, **14**, 69–72.

Schramm, A., De Beer, D., Gieseke, A., and Amann, R. (2000). Microenvironment and distribution of nitrifying bacteria in a membrane-bound biofilm. *Environ. Microbiol.*, **2**, 680–6

Seal, K. J. (1994). Test methods and standards for biodegradable plastics. In *Chemistry and Technology of Biodegradable Polymers*, Griffin, G. J. L. (Ed.). Blackie, London, pp. 116–34.

Shirakawa, M. A., Selmo, S. M., Cincotto, M. A., Gaylarde, C. C., Brazolin, S., and Gambale, W. (2002). Susceptibility of phosphogypsum to fungal growth and the effect of various biocides. *Int. Biodeterior. Biodeg.*, **49**, 293–8.

Souza, A. and Gaylarde, C. C. (2002). Biodeterioration of varnished wood with and without biocide: Implications for standard test methods. *Int. Biodeterior. Biodeg.*, **49**, 21–5.

Stender, H. Fiandaca, M., Hyldig-Nielsen, J. J., and Coull, J. (2002). PNA for rapid microbiology. *J. Microbiol. Methods*, **48**, 1–17.

Surman, S. B., Walker, J. T., Goddard, D. T., Morton, L. H. G., Keevil, C. W., Weaver, W., Skinner, A., Hanson, K., Caldwell, D., and Kurtz, J. (1996). Comparison of microscope techniques for the examination of biofilms. *J. Microbiol. Methods*, **25**, 57–70.

Tatusov, R. L., Natate, D. A., Garkavtsev, I. V., Tutusova, T. A., Shankavaram, U. T., Rao, B. S., Kiryutin, B., Galperin, M. Y., Fedorova, N. D., and Koonin, E. V. (2001). The COG database: New developments in phylogenetic classification of proteins from complete genomes. *Nucleic Acid Res.*, **29**, 22–8.

Tuomela, M., Vikman, M., Hatakka, A., and Itävaara, M. (2000). Biodegradation of lignin in a compost environment: A review. *Bioresource Technol.*, **72**, 169–83.

Urzi, C. and Albertano, P. (2001). Studying phototrophic and heterotrophic microbial communities on stone monuments. In *Microbial Growth in Biofilms, Part A: Developmental and Molecular Biological Aspects*, Vol. 336 of Academic Methods in Enzymology Series, Doyle, R. J. (Ed.). Academic Press, San Diego, CA, pp. 340–55.

Vainio, E. J. and Hantula, J. (2000). Direct analysis of wood-inhabiting fungi using denaturing gradient gel electrophoresis of amplified ribosomal DNA. *Mycol. Res.*, **104**, 927–36.

Yang, X. F., Vang, C., Tallman, D. E., Bierwagen, G. P., Croll, S. G., and Rohlik, S. (2001). Weathering degradation of a polyurethane coating. *Polym. Degrad. and Stabil.*, **74**, 341–51.

Ye, R. W., Wang, T., Bedzyk, L., and Croker, K. M. (2001). Application of DNA microarrays in microbial systems. *J. Microbiol. Methods*, **47**, 257–72.

Young, J. C. (1995). Microwave-assisted extraction of fungal metabolite ergosterol and total fatty acids. *J. Agric. Food Chem*, **43**, 2904–10.

6

The control of biodeterioration

The main effort in the field of biodeterioration has been to develop, either empirically or by design, methods for preventing the biodeterioration of materials and thus preserve their value and usefulness for as long as possible. The preceding chapters have contained information on control methods specific to a particular material, product, or structure. It is useful, however, to discuss control strategies in a wider context so that they can be applied to full effect in a given situation. The phrase that *an ounce of prevention is better than a pound of cure* is extremely pertinent to biodeterioration. As outlined in the Introduction, preventative techniques are the main option for consideration, whereas remedial treatment is at best only a temporary solution: it does not usually completely cure the problem. However, in the construction industry, a whole subindustry has built up around remedial treatment, and this often includes the use of preservatives.

Prevention should ideally commence when the raw material has assumed a potentially susceptible state. This may be directly after the harvesting of a foodstuff or when a synthetic product is placed in an environment conducive to biological activity. The first use of preventative methods was to enable the storage of foodstuffs after harvest or hunting, and these involved physical/mechanical techniques such as heat, cold, drying, osmotic pressure, and the use of mechanical barriers. Chemical methods were first introduced as fumigants (sulfur) and then as salts of mercury, copper, and zinc in the eighteenth and nineteenth centuries for preserving timber and other natural products in storage. It was not until the Second World War that the widespread use of chemical preservatives – or biocides – was seriously contemplated to prevent microbial growth on military equipment used in the tropics. Today, biocides are used routinely in the protection of materials, in the same way as agricultural pesticides are deployed for crop protection. As a result, controls on the marketing, labelling, and subsequent use of biocides have been introduced almost worldwide to protect human health and the environment.

A short review of the methods used in the prevention and control of biodeterioration will illustrate the points previously discussed. It is worth noting that no control method offers a panacea, nor is it perfect. Often we are attempting only to control the level of contamination or growth in or on a material at a minimum level acceptable to the industry and the end user. Thus different standards will apply to, for example, the pharmaceutical, food, and engineering industries.

Physical methods

A knowledge of the physiology of the biodeteriogen is useful in determining the strategy of physical methods, from the gnawing and vertical jumping capabilities of rodents, to the tolerance of certain fungi to low water activities. Refrigeration down to −20 °C is often necessary to retard growth over a long period, although slow-growing moulds have been reported in large cold stores. It is usually accepted that growth is inhibited at the temperature at which the cell contents freeze. The ability of some organisms to produce their own intracellular 'antifreeze' may account for continued growth in extreme cold. These osmolytes were first described in fungi by Nickerson and Carroll in 1945; they protect against freezing, excessive heat, and desiccation. Their presence has been recorded in all forms of life except protozoa, myxobacteria, and some simple animals. It is not uncommon for fungal growth to occur on foods stored in domestic refrigerators where the temperature may be between 4 and 6 °C. At the other end of the temperature spectrum, temperatures in excess of 40 °C will reduce the activity of a number of organisms, although temperatures above boiling point will be necessary for severe reductions in the viability of organisms which produce spores. Complete sterilization is effected only at 121 °C for a min of 15 min by a pressurized autoclave at 1.05 kgf/cm^2 (15 lb/in.^{2}). Bulky materials should be kept under these conditions for 60 min to allow for complete penetration of the heat. The exact procedure used will depend upon the thermostability of the product being sterilized. The procedure previously described is practised only in the food and pharmaceutical industries, where pathogens such as thermophilic genera of *Clostridium* and *Bacillus* must be controlled. Heating to 60 °C by the technique of 'solar bagging' is used to eliminate pest infections in museum artefacts. The materials are wrapped in black plastic sheets and exposed to the sun. The temperature rises rapidly, and adult insects, eggs, and the intermediate stages are killed by dehydration within a matter of hours. This is a low-tech method, ideal for use in hot countries, but it was developed and is used at the Canadian Conservation Institute. The use of flash heating, pasteurization, and localized increases in temperature in recirculating systems have been suggested to reduce fungal and bacterial contamination levels in oils and fuels.

The partial removal of water from a product or the maintenance of that product in an atmosphere low in moisture is commonly used in a variety of familiar situations. Desiccants are often placed inside the packaging of goods to absorb atmospheric water and reduce the humidity in the package, which in turn reduces the likelihood of microbial growth. The level of moisture in the air, often given as the RH, depends on the temperature, and both these parameters influence the expected lifetime of an object. Table 6.1 shows the length of time for which an unstable organic substrate (a photographic colour slide) may be expected to survive under different conditions of temperature and RH without any intervention from microorganisms.

Although moisture content is an easily measured and monitored parameter, it is not an absolute indicator of the potential for microbial growth on a material. The water activity (A_w) of a material is a measure of the availability of the water, in this context, to the microorganism and the atmosphere in which it grows. It is expressed as the ratio of the vapour pressure of water over the material to the vapour pressure over pure water at the same temperature. The water activity of pure water is thus 1.0, and this decreases as solutes are added. There is a direct relationship with the RH of the surrounding atmosphere, and when the system containing the material is in equilibrium at a constant temperature, we can measure the equilibrium relative humidity (ERH), which is expressed as a percentage. Thus a sample of cereal grain with a water activity of 0.75 will equilibrate with the surrounding air to produce an ERH of 75.0%. Such a figure can be related to ranges of water activity which permit the growth of microorganisms (Table 6.2). Because of the variation in chemical composition of materials, the same moisture content will not necessarily result in the same water activity. Once we have established the relationship between moisture content and water activity or ERH, moisture content can be used as a reliable measure of the safe level at which the material may be stored. There are a number of publications relating in particular to the food industry and its introduction of the 'intermediate-moisture' food; one of these is referenced for further study. Although the water activity of a material

Table 6.1. Relationship between relative humidity (RH), temperature (T), and the expected lifetime in years of a photographic slide

RH\T %	6 °C	8 °C	11 °C	14 °C	17 °C	19 °C	22 °C	25 °C	28 °C	31 °C	33 °C
30	525	356	243	168	116	81	57	40	29	21	15
35	451	307	210	145	101	71	50	35	25	18	13
40	387	264	182	126	88	62	43	31	22	16	12
45	333	228	157	109	76	54	38	27	19	14	10
50	287	197	136	95	66	47	33	24	17	12	9
55	247	170	118	82	58	41	29	21	15	11	8
60	213	147	102	72	51	36	26	18	13	10	7
65	184	128	89	62	44	31	22	16	12	9	6
70	160	111	77	54	39	28	20	14	10	8	6
75	138	96	67	48	34	24	17	13	9	7	5
80	120	84	59	42	30	21	15	11	8	6	4
85	104	73	51	36	26	19	14	10	7	6	4
90	90	63	45	32	23	16	12	9	6	5	3
95	79	55	39	28	20	15	11	8	6	4	3

We are grateful to Dr Saulo Guths for providing this table.

Table 6.2. Threshold water activities for the growth of microorganisms

Microbial group	A_W
Normal bacteria	0.91
Normal yeasts	0.88
Normal fungi	0.80
Halophilic bacteria	0.75
Xerophilic fungi	0.65
Osmophilic yeasts	0.60

or substance is an intrinsic feature, preservation methods often rely on altering the ERH of a product by the addition of chemicals which change the water relations within that product. Propylene glycol and sorbitol are used in tobaccos and domestic animal food to impart a feeling of moisture or plasticizing effect on the product. These glycols also reduce the water activity sufficiently to retard the growth of moulds. Water activity is also reduced by use of inorganic salts and sugars; thus the use of brine pickling for foods and sugar in jams (known as preserves!).

Radiations such as gamma rays, UV rays, and microwaves have all been employed to a limited extent as sterilizing agents, often where there has been a large-scale spoilage problem in, for example, a liquid product, and there is a possibility that the product may be reclaimed or reworked. Gamma irradiation has been used to treat books from a library after flooding resulted in widespread fungal growth. The ability of the radiation to adequately penetrate the material to be sterilized is paramount. Thus UV systems have found a niche in the treatment of both potable and recirculating water used in industry, in which relatively thin films can be made to flow through the source of radiation. Even so, algal growth can develop or dirt can build up on the surface of the glass tubes through which the water flows, preventing the sterilization process. The residual preservative activity of irradiation is nil, and other precautions must be subsequently used to prevent reinfection.

Filters are also extensively used as barriers to microorganisms in aqueous recirculation systems. Membrane filters with pore sizes of either 0.45 or 0.22 μm are routinely used to trap fungi and bacteria. Larger-pore filters can be employed for filamentous growths to reduce blockage problems. The use of physical barriers for the exclusion of larger organisms is a matter of common sense and knowledge of the habits of particular troublesome groups. This area has been adequately discussed in Chapters 2 and 4.

The nature of the surface of a material will affect the initial colonization by microorganisms. Smooth hydrophobic surfaces such as epoxy paint films are less likely to encourage colonization than are textured surfaces (external renderings or stucco plasters), which can harbour organic debris and water and provide crevices for microorganisms. A novel application of this approach has been devised for the prevention of external fouling on off-shore structures. It involves coating (200 μm thickness) the vulnerable parts of the structure with a silicone rubber material (polydimethyl siloxanes are quoted as being the best class of compound) which contains a hydrophobic fluid. The fluid slowly and continuously exudes from the surface, preventing settlement of marine organisms. Field trials have shown that it can significantly reduce fouling for up to 4 years. Self-polishing, antifouling paints on the hulls of ships are another example.

Chemical methods

The chemicals used in controlling biodeterioration in materials come in the form of gases (fumigants), dispersable powders, and liquids. There is some overlap, in that similar compounds may be formulated either for particulate dispersion as a fumigant or for use as a wettable powder in the control of insect biodeteriogens. Likewise, some compounds, for example formaldehyde, although employed as liquid preparations, exert their effect in the gaseous phase when incorporated into a product.

Gaseous sterilization or fumigation is used to decontaminate materials which have already been infested with insects or microorganisms and which are not amenable to other forms of sterilization such as heat, radiation, or the addition of toxic chemicals in solution or suspension. Their successful use requires specialist knowledge and equipment to cope with the attendant dangers in handling the sterilants. Methyl bromide has been used for insect pests for many years, but will be phased out in many countries by 2004 and will eventually be withdrawn also in developing countries. Ethylene oxide and propylene oxide are still used for bacteria and fungi control in some countries (ethylene oxide is banned in the UK except for some specialist medical applications). Their advantage lies in their ability to penetrate packed materials such as bales of tobacco. The materials are placed in chambers in which a vacuum is drawn before introduction of the sterilant (at defined concentration, temperature, and humidity) to encourage good penetration. It is, of course, important that all traces of the sterilant are removed from the chamber before it is reopened and from the sterilized product; materials treated with ethylene oxide can release gas long after treatment if procedures are not totally satisfactory. Safety procedures must be employed to ensure there is no hazard to personnel.

Low oxygen atmospheres (less than 0.03%) have been used to treat infestations of woodborers and termites in museum artefacts. In this case, there is little danger to personnel, but rigid control of temperature and RH are essential if treatment is to be successful. Hence trained operators are necessary.

Chemical preservatives are variously referred to as biocides, bactericides, fungicides, fungistats, antifouling compounds, and material protectants. The term *preservative* is probably the most accurate descriptive term for general use, because it infers that the protected material maintains its integrity and performance characteristics during storage and use. The term *biocide* is more commonly used to describe the range of chemicals used to combat biodeterioration in industrial products, although we still talk of wood preservatives and antifouling compounds in the building and boat industries, respectively. We must thus expect to encounter different terms in describing the same chemical. For brevity, the term biocide is used in the following discussions.

To be effective, the ideal biocide (which does not, of course, exist!) should have the characteristics described in the following subsections.

TOXIC TO THE TARGET DETERIOGENS Some biocides claim a wide spectrum of activity, covering both fungi and bacteria, whereas others are more specific to fungi or bacteria. Different chemical species will also be used against the algae, insects, arthropods, and rodents. Rodent control is discussed more fully in Chapter 4. It is thus important to identify the causative organisms so that the correct type of biocide is used.

NON-TOXIC TO HUMAN AND NON-TARGET ANIMAL AND PLANT LIFE By their very nature, biocides are toxic products, and it is very difficult to reconcile the two requirements. However, in recent years products have come onto the market with high LD_{50} values (up to 10 000 mg/kg oral dose in rats), making them more environmentally acceptable. The biodegradability of a biocide which may leach out from a product after application, or any waste biocide remaining after processing or in a discarded liquid such as a metal-working emulsion, must be considered. In this instance, the biocide is in its diluted form and may be in a partially inactivated or degraded state. However, it may still be capable of affecting the operation of a sewage treatment plant or adversely affecting the soil microflora, unless further dilution occurs. National and international regulatory controls, for example, the U.S. Federal Insecticide, Fungicide, and Rodenticide Act (FIFRA) and the European Union BPD, are now in place to enable hazard- and risk-related assessments of toxic chemicals and thus restrict their exposure to humans and the environment. This has resulted in the introduction of less toxic biocidal actives which, under controlled use, present no threat to the environment.

IT MUST BE COMPATIBLE WITH THE PRODUCT AND NOT IMPART ANY UNWANTED COLOUR OR ALTER THE PROPERTIES OF THE MATERIAL It should not adversely affect any coating which may be applied (copper-containing compounds can impart a green or blue stain, and chelating biocides will react with metals to produce a coloured pigment), nor should it corrode the product or its container. Metal-containing compounds, for example, cannot be used in aircraft jet fuels, because they deposit the metal on the surfaces of the turbine blades as the fuel is burnt. Phenolics may hasten the chemical degradation of cellulose in strong sunlight (actinic tendering) whereas divalent inorganic salts used in the stabilization of isothiazolinones can cause coagulation in some polymer lattices.

IT MUST PARTITION INTO THE PHASE WHERE ITS ACTIVITY IS REQUIRED. In practice, all biocides must exhibit some aqueous solubility. This is important in oil–water emulsions, in which a low partition in the water phase would prevent any biocidal activity where it is required. Cosmetic preparations often comprise oil-in-water, water-in-oil, or oil-in-water-in-oil, and this may make them very difficult to preserve, as the partitioning of the preservative is critical. Solvent-based products, such as paints and plastics, ideally require a biocide with solubility in non-polar solvents, so that an even dispersion is effected during manufacture. A low water solubility will also help in reducing the leach rate when water is present, and this will make the biocide more effective and longer lasting.

IT SHOULD BE STABLE IN BOTH THE CONCENTRATE AND THE DILUTED FORM The effects of the external or product environment should not deactivate the biocide during its working life. Stability assumes a greater importance in materials to which the biocide can be added only during manufacture and one dose must suffice. Many biocidal actives in use today have to strike a fine balance between efficacy as a preservative and toxicity/environmental fate. As a result, they are often destabilized by adverse pH, temperature, light, and reactive chemicals in products such as reducing agents, surfactants, and proteins. The requirement for 12 to 24 month shelf lives for products such as cosmetics puts a strain on the preservation system in such circumstances. In other situations, such as recirculating aqueous systems, biocide can be added as a top-up or tank-side additive when its activity decreases.

IT SHOULD HAVE A LOW COST:PERFORMANCE RATIO This generally means that it should exhibit its effect economically at low dose levels and preserve the material for its working life.

MAJOR CHEMICAL ACTIVES

OXIDIZING AGENTS The biocide with the longest history in water disinfection, and still one of the most popular antimicrobials because of its low cost, is chlorine. It is used in industrial and domestic applications. However, it has a limited pH range, being less effective at alkaline levels, and is inactivated by reaction with proteins, as well as with ammonia-containing compounds. The latter reaction may result in the formation of carcinogenic trichloromethanes. Various alternatives have been explored, including bleach, bleach with bromide, chloramines, bromochlorodimethyl hydantoin (BCDMH), ozone, and chlorine dioxide. Chlorine dioxide offers the advantages of selectivity, effectiveness over a wide pH range, and speed of kill, but safety and cost issues have so far restricted its use.

Other oxidizing biocides include hydrogen peroxide, ozone, and, of course, other halogens. Ozone is being increasingly used in place of chlorine in the treatment of domestic and industrial water supplies. It is generated on site, removing any transport problems, and decomposes very rapidly, eliminating any residual effects on the environment. However, its biocidal activity is much reduced at high pH values, and it is reactive with a wide range of compounds, leading to its inactivation.

ALDEHYDES Formaldehyde and glutaraldehyde are broad-spectrum biocides, with good sporicidal activity. They have good water solubility, but their main feature is that they easily vaporize, and this is put to good effect in situations in which there is poor hygiene in the factory where the product is made. Formaldehyde, in particular, readily reacts with a range of glycols or amines to produce O-formals and N-formals, respectively. These are known as formaldehyde condensates or formaldehyde donors, and they slowly hydrolyse to release formaldehyde in the product or process. If added (usually in combination with another active) early on in the manufacturing process, they provide protection throughout the plant and then continue to release formaldehyde during storage of the finished product. The rate at which formaldehyde is released is determined by the type of formal (O-formals tend to be less stable than N-formals) and the equilibrium conditions in the product. The restriction of formaldehyde vapour levels in the workplace because of its sensitizing and carcinogenic properties has resulted in the development of donors with low free formaldehyde levels (typically <1%). Free formaldehyde is the released amount as defined by a specific analytical technique, for example, the Hantzsch method. Formaldehyde donors have been used successfully for many years in the MWF industry, in which their enhanced activity at alkaline pH and in the vapour phase gives them distinct advantages over other actives. They continue also to be used in combination with isothiazolinones as in-can biocides for paints and adhesives and in combination with parabens (see the subsection on Isothiazolinones) in cosmetic preparations.

Glutaraldehyde is used as a disinfectant for the cold sterilization of surgical instruments and endoscopes and as an alternative to heat treatment in an autoclave. However, the effective dose level is between 2% and 3% and an exposure time of at least 10 h is needed. It is also a sensitizer and is monitored in the workplace to avoid excessive exposure. As a disinfectant it is used at lower levels (approximately 0.15%) in wet wipes and sprays. It is employed at high levels in the oil industry as a cost-effective disinfectant for large volumes of water pumped into wells for secondary oil extraction.

ALCOHOLS The aliphatic alcohols, such as ethanol and isopropanol, tend to be used as disinfectants in antibacterial hand washes because of their speed of action in the destruction of bacteria and viruses. Because of their evaporation rate they cannot be used as biocides. However, benzyl alcohol and phenoxyethanol are used in cosmetic preservation as bactericides at concentrations up to 1%, often in combination with complementary actives.

PHENOLICS Phenolic compounds, derived from coal tar, were the early effective disinfectants. Phenol itself is a potent carcinogen, but a range of chlorinated phenols is still available for both disinfection and preservation. They tend to be bactericides, but orthophenyl phenol has fungicidal activity, and its potassium salt is often used in combination with a quaternary ammonium compound to provide a broad-spectrum biocidal wash for the treatment of masonry before repainting. Some compounds have a characteristic odour which limits their use, and the polychlorinated phenols (e.g., trichlorophenol) have mostly been abandoned because of their recalcitrance and persistence in the environment. On the whole, this class of compound is regarded as 'old technology' in today's world of preservation.

ORGANIC ACIDS AND THEIR ESTERS Acetic, propionic, lactic, sorbic, and benzoic acids and their salts are traditionally used for preservation of acidic foodstuffs. They are weak acids which are active only in their undissociated states, thus limiting their effectiveness to the pH range 4–6. At this range they are employed to control yeast and mould growth in fruit juices and fermented milk products. The esters of hydoxybenzoic acid, known collectively as the parabens, are very widely used in the food and cosmetic industries as preservatives to inhibit moulds and yeasts. Their advantages over the acids are their extended pH range from 4 to 8.5 and their synergistic activities when used in combinations. As the chain length of the alkyl group increases, the water solubility decreases and the antimicrobial activity increases. Three (methyl, ethyl and propyl) and five (methyl, ethyl, propyl, butyl, and isobutyl) paraben combinations are commonly used for the protection of leave-on cosmetics.

QUATERNARY AMMONIUM/PHOSPHONIUM COMPOUNDS The activity of this relatively large class of antimicrobials results from their cationic surfactant nature, although anionic compounds are also known, depending on the nature of the hydrophobic portion of these amphipathic molecules. Their antimicrobial activity increases with lengthening side chain, optimal efficacy being at C14–C16. They are rapid-acting antimicrobials and thus find application as disinfectants. However, because of their stability to pH and heat, they can be successful as longer-term biocides, providing they are not used in combination with anionic surfactants or high levels of protein and salts, which will severely reduce their efficacy. They have broad-spectrum activity towards bacteria and will control algal growth. Because of their cationic nature, they show some substantivity to negatively charged surfaces such as cellulose. In this respect they will control sapstain on freshly felled wood and prevent bacterial growth on cotton textiles. The 'quats', as they are known, are generally regarded as environmentally friendly biocides, apart from their inclination to produce foaming when used at high concentrations.

ISOTHIAZOLINONES This is a group of actives which might be termed 'modern technology' in the biocide field. They are characterized by the five-membered isothiazolinone ring, which has a number of different substituents which give the resulting molecule varying biocidal properties, including, generally, good activity against microorganisms in biofilms. The main derivatives in commercial use are methyl, methylchloro, benzyl, octyl, and dichloro-octyl. The methylchloro (CIT) derivative is an extremely cost-effective biocide which controls a wide range of bacteria and fungi at levels in the range 2–10 ppm. It is water soluble and used widely as an in-can preservative. The methyl (MIT) and benzyl (BIT) derivatives tend to have better bactericidal activity and are used in the range 50–200 ppm for in-can protection, whereas the octyl (OIT) and dichloro-octyl (DCOIT) derivatives have low water solubility and good intrinsic fungal/algal activity. Because of this, they are employed as dry-film preservatives in products such as paints, adhesives, sealants, and plastic film. They are progressively deactivated at alkaline pH (>8.5) and temperatures in excess of 60 °C, as well as in the presence of nucleophilic agents.

OTHER ORGANOHALOGEN-CONTAINING BIOCIDES The other major preservative used for in-can preservation is bromonitropropane diol (trade name, Bronopol). Bronopol shares the same deactivating characteristics as the isothiazolinones, but it is more effective as a bactericide than a fungicide. It is often used in combination with the water-soluble isothiazolinones to provide an extremely broad-spectrum preservative.

Several other important biocides are worth a mention as they provide dry-film protection against either algal or fungal defacement on surfaces which are often decorative in nature. Carbendazim is a fungicide which has its roots in the agrochemical sector. It has a very low water solubility, is very durable to external environments, and has an excellent antifungal profile. Although there are some gaps in its spectrum of activity (*Alternaria* sp. being a notable example), in combination with other fungicides, for example, octylisothiazolinone, it has been shown to provide long-term protection to external coatings on buildings. Iodopropynyl butylcarbamate (IPBC) was developed as an antisapstain agent for wood but is now used in exterior paints and was recently approved as a cosmetic preservative in Europe.

For the prevention of algal growth on exterior coatings applied to buildings and structures, the biocide industry has used agricultural herbicides, the most extensively used molecule being dichlorophenyl dimethylurea (trade name, Diuron). This is an extremely durable algicide, and, in combination with Carbendazim and octylisothiazolinone, has provided the coatings industry with the most effective dry-film protection system for the past 20 years.

Examples of the major actives and their applications are listed in Table 6.3, and the reader is referred to the reference list relating to this chapter, where further information on the actives and their properties can be found.

Modes of action

Biocides exert their effect on the target organism in a number of ways, including oxidation, hydrolysis, denaturation, cell lysis, metabolic inhibition, and alteration of membrane permeability (Table 6.4). Their efficacies, often quoted as minimum inhibitory concentrations, will vary from less than 10 ppm up to several thousand (Table 6.5). In use their activities are affected by both the ingredients in the product and the environment in which they have to work. For example, concentrations of biocide between three and ten times the normal levels for control of planktonic (free-floating) bacteria have been quoted as necessary to kill those organisms associated with a biofilm. It has been suggested that the glycocalyx, formed by a polymeric matrix of polysaccharides secreted by the bacteria in the film, acts as an ion-exchange resin, limiting the penetration of charged molecules. This may play a role in controlling the biocide activity as well as the altered physiological state of sessile cells. It has been shown that preventing the formation of a normal biofilm, by manipulating the quorum sensing behaviour of the bacterial cells, results in increased sensitivity to sodium dodecyl sulfate. Future research in this area may help us to destroy deteriogenic biofilms without increasing biocide levels.

No single biocide in use today is capable of solving all biodeterioration problems, and it is important to know their limitations to ensure that the right system is used. Table 6.6 gives a summary of the general spectrum of activity of the major biocide groups and some of the more common blends of biocide actives are listed in Table 6.7.

Table 6.3. Major biocidal actives and their application areas

Applications:
1. Adhesives
2. Antifouling
3. Cosmetics and pharmaceuticals
4. Disinfectants and antiseptics
5. Foodstuffs
6. Fuels
7. Leather
8. MWFs
9. Paints, in-can, and polymer emulsions
10. Paints, dry film
11. Paper and pulp
12. Plasters
13. Plastics
14. Printing inks and lithographic solutions
15. Process water treatment
16. Sealants, putties, and grouts
17. Surfactants
18. Textiles
19. Wood

Note: This list is not exhaustive either with respect to the actives listed, their applications, or their regulatory status. The CAS numbers are given to enable correct chemical identification of the actives.

Trivial name	CAS no.	1	2	3	4	5	6	7	8	9	10	11	12	13	14	15	16	17	18	19
Benzalkonium chloride	63449-41-2			+	+			+												+
Benzethonium chloride	121-54-0			+	+														+	
Benzoic acid and its salts	65-85-0			+		+														
Benzoic acid and its salts	65-85-0			+		+														
Benzyl alcohol	100-51-6			+																
Bis oxazolidine	66404-44-2	+					+		+											
Bis morpholinomethane	5625-90-1								+											

216

Compound	CAS number											
Benzisothiazolinone	2634-33-5	+					+		+	+		
Bronidox	30007-47-7	+				+	+		+	+		
Bronopol	52-51-7	+				+	+		+	+		
Carbendazim	10605-21-7			+		+			+			+
Cetrimide	57-09-0		+	+								
Chlorthalonil	19897-45-6			+	+							
Chloroallyl methenamine chloride	51229-78-8	+			+	+			+	+		
Chloromethyl isothiazolinone	26172-55-4	+			+	+			+	+		+
Copper naphthenate	1338-02-9		+				+					+
Dazomet	533-74-4	+					+					
Dehydroacetic acid	520-45-6	+								+		
Diazolidinyl urea	78491-02-8	+					+		+	+		
Dibromodicyanobutane	35691-65-7	+				+			+			
Dibromonitrilopropionamide	10222-01-2	+						+	+			
Dichlorooctylisothiazolinone	64359-81-5	+			+	+		+	+	+		+
Dichlorofluanide	1085-98-9					+				+	+	
Dichlorophen	97-23-4	+			+		+	+				
Diuron	330-54-1		+				+		+			
DMDM hydantoin	6440-58-0	+		+								+
Fluorofolpet	719-96-0					+						
Formaldehyde and its condensates	50-00-0	+		+	+	+			+	+		

(continued)

217

Table 6.3. *(continued)*

Trivial name	CAS no.	1	2	3	4	5	6	7	8	9	10	11	12	13	14	15	16	17	18	19
Glutaraldehyde	111-30-8			+	+													+		+
Hexahydrotriazine	4719-04-4						+	+				+								
Imazalil	35554-44-0								+											+
Imidazolidinyl urea	118215-45-5			+																
IPBC	55406-53-6			+			+			+			+		+	+	+			+
Irgarol	28159-98-0									+			+							
Methylene bisthiocyanate	6317-18-6								+							+				
Methylisothiazolinone	2682-20-4	+		+						+			+		+	+			+	
Oxybisphenoxyarsine	58-36-6			+						+					+					
Octylisothiazolinone	26530-20-1			+			+	+	+	+			+	+	+	+			+	+
Orthophenyl phenol	90-43-7	+		+	+			+	+			+		+					+	+
Paraben esters, e.g., methyl paraben	99-76-3					+														
Parachlorometacresol	59-50-7	+		+				+												
Parachlorometaxylenol	88-04-0				+															
Pentachlorophenol	87-86-5							+									+	+		
Pentachlorophenol laurate	3772-94-9				+														+	
Peracetic acid	79-21-0		+																	
Phenoxyethanol	122-99-6		+					+										+		
Polyhexamethylene biguanide	35708-79-3		+																+	

Propiconazole	60207-90-1								+
Salicylic acid and its salts	69-72-7	+					+		
Silver and its salts	7440-22-4	+	+						
Sodium pyrithione	3811-73-2			+					
Sorbic acid and its salts	110-44-1	+	+						
Tributyltin esters, e.g., oxide	56-35-9	+						+	
Triclocarban	101-20-2	+				+			
TCMBT	21564-17-0			+				+	+
Tebuconazole	107534-96-3				+				
Thiomersal	54-64-8	+			+	+			
Triclosan	3380-34-5	+		+	+			+	
Zinc oxide		+	+						
Zinc pyrithione	13463-41-7	+	+	+	+	+	+	+	

DMDM = Dimethyl dimethylol.
TCMBT = 2, thiocyanomethylthiobenzothiazole

219

Table 6.4. Mechanisms of action of biocides

Mode of action	Examples
Oxidising, causes lysis of cell wall and cell constituents	Hypochlorite, bromine and its compounds, ozone
Membrane active, may affect active transport mechanisms or disrupt membrane integrity	Quaternary ammonium compounds, alcohols, parabens
Protein denaturation	Phenols, aldehydes, formaldehyde condensates, and parabens
Protein synthesis inhibitor, bind with thiol groups in cell affecting enzyme activity	Isothiazolinones, Bronopol, dibromodicyanobutane
Nuclear division inhibitor, DNA synthesis inhibition	IPBC, Carbendazim
Membrane synthesis inhibitor, prevents the synthesis of ergosterol in fungi	Imazalil, Tebuconazole, Propiconazole
Photosynthesis inhibition, affects electron transport	Diuron, Irgarol, Terbutryn

In recent years, a new approach to the use of chemicals for the control of biodeteriogens has emerged in order to improve both the environmental problems associated with toxic chemicals and to improve the efficacy of such chemicals over longer periods. This has been the introduction of controlled-release systems, in which the toxic agent is held in a suitable matrix and is slowly released with time, usually in an aqueous environment. A range of matrices has been used, including a number of polymeric elastomers such as natural rubber, chloroprene, polyurethane, polybutadiene, and nitrile rubber. The chemical may simply leach out, by means of a migration mechanism, or be released as the result of the hydrolysis of a polymer/toxin ester (e.g., tributyltin acrylate or methacrylate copolymerized with a film-forming polymer such as PVC). The release rate may be adjusted by the use of water-soluble fillers (accelerators) or water-insoluble organic compounds (retardants). The main areas of application are in antifouling, although systems have been suggested for fungal and bacterial control in other situations.

The problem inherent in these systems is controlling the release rate with any accuracy over long periods. There has really been an impetus to do this only in the drug industry, in which controlled-drug-release systems

Table 6.5. Comparative minimum inhibitory concentrations (ppm) for selected biocide actives

Active	Gram-negative bacteria	Gram-positive bacteria	Fungi	Algae
Phenoxyethanol	4000	2000	4000	ND
Methyl paraben	3000	2000	1000	ND
Benzalkonium chloride	200	10	50	1
Tebuconazole	>5000	>5000	50	ND
Formaldehyde	60	15	500	ND
CIT/MIT	5	5	8	10
MIT	40	100	250	ND
BIT	100	40	350	ND
Bronopol	50	100	1000	
IPBC	>1000	>1000	5	ND
OIT	500	20	5	5
Zinc pyrithione	400	10	50	ND
Carbendazim	>1000	>1000	0.5	ND
Diuron	>5000	>5000	>1000	1

Note: The minimum inhibitory concentrations quoted are averages of data from the literature determined by a variety of methods and do not necessarily predict performance in a product.
ND, no data available.

are effective because of their relative short life (a few days as opposed to years) and the controlled environment in which they are active. Although some crude control is probably possible over a 2–3-year period, it would not be realistic to say that biocide release was being actively controlled or that systems could be finely tuned to meet the requirements of a particular situation. Such a hope, along with the development of biocides which can be targeted at specific groups of biodeteriogens, must be reserved for the future.

REGULATORY ASPECTS

The majority of biocidal actives are termed toxic chemicals and as such are subject to national and international controls. These controls range from inventories, or lists of permitted and banned chemicals, through hazard and risk assessments which control the marketing of actives, to the use of risk labels and maximum permitted levels of use. Probably the two most

Table 6.6. Biocide activity spectra

Group	Examples	Bacteria	Fungi	Algae
Phenols	o-phenylphenol, p-chloro-m-cresol	+	+/−	−
Organic acids/esters	Benzoic acid, methyl paraben	+/−	+	−
Quaternary ammonium compounds	Benzalkonium chloride, PHMB	+(p)	+	+/−
Aldehydes	Formaldehyde and its condensates	+	−	−
Heterocyclic N,S compounds	Isothiazolinones	+	+	+/−
Organohalogens	Bronopol	+	+	−
Pyridine derivatives	Zinc/sodium pyrithione	−	+	−
Azoles	Tebuconazole, Propiconazole	−	+	−
Carbamates	IPBC, Carbendazim	−	+	−
Amides/ureas	DBNPA, Diuron	+/−	−	+/−
Triazines	Hexahydrotriazine, Terbutryn	+/−	−	+/−

+, active; −, weak to no activity; +/−, variable depending on compound; p, active predominantly against gram-positive bacteria. PHMB, polyhexamethylene biguanide; DBNPA, Dibromonitrilopropionamide.
Note: The preceding data represent the predominant activities of the various biocide groups and show their weaknesses in any preservation system. Advice on the activities of all preservative systems should first be obtained from the suppliers before implementation.

significant pieces of legislation which control the use of biocidal actives are the Federal Insecticide, Fungicide, and Rodenticide Act (FIFRA) and the Biocidal Products Directive (BPD). FIFRA was introduced in the United States some 20 years ago to control the use of pesticides and thus covers both agricultural and industrial applications. The U.S. Environmental Protection Agency administers the requirements of the Act, the most important aspect of which is the requirement for all pesticides to be registered. The BPD is similar to FIFRA but supplements the Plant Protection Products Directive and covers only non-agricultural pesticide actives. It was introduced in 2002 and is a European Union Directive which must be integrated into member states' legislation. It is scheduled to take 10 years for complete registration of all actives and formulations, but this is thought to be slightly optimistic. The registration schemes for both involve the submission of a dossier of toxicological and environmental fate data

---

Table 6.7. Some typical biocide active combinations

Actives	Applications
CIT/MIT/HCHO condensate	In-can preservation of paints & other aqueous emulsion systems
CIT/MIT/Bronopol	As above but where HCHO use is restricted
BIT/HCHO condensate	In-can aqueous preservation in high pH systems
HCHO condensate/IPBC	Metal-working fluids
BAC/K o-phenylphenate	Biocidal washes
MIT/BIT	In-can aqueous preservation of alkaline systems & where CIT use restricted
Diuron/OIT/Carbendazim	Algal/fungal protection for external coatings
OIT/Carbendazim	Anti-fungal paints in bathrooms and other high humidity interior situations
DMDM Hydantoin/IPBC	Personal care products
OIT/DCOIT	Plasticised PVC film

Note: The preceding list is far from exhaustive and only serves to illustrate how actives may be combined to extend their individual spectrum of activity

---

(the base set), together with additional data dependent on the application. The data are reviewed, the actives placed on a positive list, and hazard phrases are assigned from which a risk assessment can be performed to define how each active may be used. All states must observe a mutual recognition of the registration process. The purpose of both schemes is to protect human health and the environment by controlling exposure to a safe limit. In practice this means that there will be use limits set for biocidal actives and some will be withdrawn from use. This has already been implemented in the EU Cosmetics Directive and the Japanese equivalent scheme, where there is a positive list of permitted preservatives with both use limits and application restrictions. For example, chloroisothiazolinone/methylisothiazolinone (CIT/MIT) is limited to 15 ppm active for all applications in the EU, whereas in Japan it is restricted to rinse-off use only. As more controls are implemented, the need to adequately preserve products will result in more multicomponent preservative combinations.

Biological methods

The use of one biological agent to suppress another is a strategy which receives periodic attention, particularly when it fails to achieve the desired result and the balance of nature is disturbed. One example in the control of wood decay is the use of immunizing commensals. Studies have shown that the fungus *Scytalidium lignicola* is antagonistic to *Lentinus lepideus*, a wood-decaying fungus; *Trichoderma viride* has also been used commercially. It may be introduced as a slug (spores of the fungus mixed in a dehydrated nutrient matrix and pressed into a pill) into the trunk of a standing tree; in a similar situation, the technique has also been suggested for protecting electricity transmission and telegraph poles. As biotechnological research develops as a discipline, the search for microbially produced chemical agents with highly specific antagonistic properties may see a resurgence, particularly in the control of contamination of fermentation processes.

WEBSITES

http://www.hse.gov.uk/hthdir/noframes/biobkgnd.htm
http://www.pestlaw.com/x/international/EC-20010430A.html (Biocidal Products
    Directive)

REFERENCES

Alexander, B. R. (2002). An assessment of the comparative sensitization of some common isothiazolinones. *Contact Dermatitis*, **46**, 191–6.
Ash, M. and Ash, I. (1996). The index of antimicrobials. Gower, Aldershot, England.
Chapman, J. S. (2003). Biocide resistance mechanisms. *Int. Biodeterior. Biodeg.* **51**, 133–8.
Cooke, M. (2002). European review of biocides. *Pharmachem.*, **1**, 48–50.
Davis, R., Birch, G. G. and Parker, K. J. (1976) Intermediate Moisture Foods. Applied Science Publishers, London.
Denyer, S. P. (1995). Mechanisms of actions of anti-bacterial biocides. *Int. Biodeterior.*, **36**, 227–45.
Denyer, S. P. and Hugo, W. B. (Eds.). (1991). *Mechanisms of Action of Chemical Biocides. Their Study and Exploitation.* Blackwell Scientific, Oxford, England.
Gaylarde, C. C. (1995). Design, selection and use of biocides. In *Bioextraction and Biodeterioration of Metals*, Gaylarde, C. C., Videla, H. A. (Eds.). Cambridge University Press, Cambridge, England, pp. 327–60.
Gillatt, J. (1993). The use of isothiazolinone-based biocides for the microbiological protection of water borne paint formulations. *Polym. Paint. Colour J.*, 12 May.
Gillatt, J. W. (1996). The use of biocides and fungicides in wood coatings and preservatives. *Pigment & Resin Technol.*, **25**(5), 4–10.
Gillatt, J. W. (1997). The effect of redox chemistry on the efficacy of biocides in polymer emulsions. *Surf. Coat. Int.*, **80**, 172–7.

Hardy, W. O. (1997). *An Introduction to Chlorine Dioxide*. Engelhard Corp., North Kingstown, RI.

Gill, C. O. and Lowry, P. D. (1982). Growth at sub-zero temperatures of black spot fungi from meat. *J. Appl. Bacteriol.*, **52**, 245–50.

Morton, L. H. G., Greenway, D. L. A., Gaylarde, C. C., and Surman, S. B. (1998). Consideration of some implications of the resistance of biofilms to biocides. *Int. Biodeterior. Biodeg.*, **41**, 247–59.

Mutasa, E. S. and Magan, N. (1990). Utilisation of potassium sorbate by tobacco spoilage fungi. *Mycol. Res.*, **94**, 965–70.

Mutasa, E. S., Seal, K. J., and Magan, N. (1990). The water content/water activity relationship of cured tobacco and water relations of associated spoilage fungi. *Int. Biodeterior.*, **26**, 381–96.

Mutasa, E. S., Magan, N., and Seal, K. J. (1990). Effects of potassium sorbate and environmental factors on the growth of tobacco spoilage fungi. *Mycol. Res.*, **94**, 971–8.

Nedwell, D. B. (1999). Effect of low temperature on microbial growth. *FEMS Microbiol. Ecol.*, **30**, 101–11.

Paulus, W. (1993). *Microbicides for the Protection of Materials*. Chapman & Hall, London.

Rogers, P. B. (1989). Potential of biocontrol organisms as a source of antifungal compounds for agrochemical and pharmaceutical development. *Pesticide Sci.*, **27**, 155–64.

Rossmore, H. (Ed.). (1995). Disinfection, sterilization and preservation problems and practices. *Int. Biodeterior.*, **36**, 195–472.

Seal, K. J. (2002). Maximising your assets – preservative blends. *Soap, Perfum. Cosmet.*, **77**(9), 61–6.

Seal, K. J. and Edge, M. (2001). The enhanced performance of preservatives for the protection of paints and coatings using a novel encapsulation process. *Farg och Lack*, **47**(1), 6–10.

Strang, T. K. (1995). The effect of thermal methods of pest control on museum collections. In *Biodeterioration of Cultural Property*, 3 ed. Aranyanak, C. and Singhasiri, C. (Eds.). Thammasat University Press, Bangkok.

Steinberg, D. (1996). *Preservatives for Cosmetics*. Allured Publishing, Carol Stream, IL.

Tadros, T. F. (1989). Colloidal aspects of pesticidal and pharmaceutical formulations – an overview. *Pesticide Sci.*, **26**, 51–77.

General Index

Aesthetic biodeterioration – definition, 3
Acetanilides utilised by *Penicillium* sp., 88
Acetic acid, 213
Acid mine waters, 95
Adhesives and sealants, 33, 101–102
Adhesive failure, associated organisms, 102
Aflatoxins, 22–23
Air conditioners, 119, 123
Aircraft, 156–158
Aircraft fuel tank liners, 50
Alcohols, as biocides, 212
Aldehydes, 211
Allergies, 140
Allergies to house mites, 26
Alpha-chloralose, 136
Aluminium rolling, 59
Amylases, 14
Animal glue, 16, 33, 124
Anti-coagulant rodenticides, 136–138
Anti-fouling paints, 83, 158
Anti-icing agents in aviation fuel, 48
Audits, plant/factory, 166–171
Aviation fuels, 48

Benzoic acid, 213
Benzothiazoles, 100
Biochemical assimilatory biodeterioration – definition, 5
Biochemical dissimilatory biodeterioration – definition, 5
Biodeterioration – definition, 1
Biodeterioration, standard test methods, 188
Biodegradation – definition, 1
Biocides, activity spectra, 222

Biocides, active combinations, 223
Biocides, controlled release systems, 220
Biocides, ideal properties, 209–211
Biocides, major actives used in industrial applications, 216–219
Biocides, minimum inhibitory concentrations (MICs), 221
Biocides, modes of action, 215–221
Biocides, preliminary screening, 178
Biocides, regulatory aspects, 221
Biocorrosion (see microbially-influenced corrosion)
Biofilms, 2–5, 50, 53, 90, 116, 119, 168
Biofilms, sampling methods, 173
Bioluminescence technique for ATP, 175
Biosurfactants, 44
Bird nests, 31, 32
Bird strikes on aircraft, 156
Bisthiocyanates, 100
Bitumen adhesives, 101
Bleach (see sodium hypochlorite)
Blue stain of wood, 15
Books, 16, 207
Bromadiolone, 137
Brodifacoum, 137
Bromethalin, 136
Bromo nitropropane diol (Bronopol), 214
Brown rot of wood, 14
Buildings (built environment), 111–152
Buildings, algae and cyanobacteria affecting structure, 118–121
Buildings, bat problems, 141
Buildings, bird problems, 122–123
Buildings, control of microbial growth, 121
Buildings, control of termites, 129

White rot of wood, 14
Wood in marine environment, 17–19
Wood, control of woodboring beetles, 124–126
Wood decay, 14, 15, 112–114, 160

Wood, insect attack, 124–130
Wool, 31

Zearalenone, 22
Zinc phosphide, 136

Organism Index

Note: Only genera and species are *italicised*.